BASIC

ENGINEERING

DRAWING

BASIC ENGINEERING DRAWING

R S RHODES

M Weld I, Diploma in Advanced Studies in Education.
Lecturer responsible for Engineering Drawing
Stafford College of Further Education.

L B COOK

B A (Hons), MIED, Cert Ed.
Lecturer in Engineering Drawing
Stafford College of Further Education.

Longman Scientific & Technical
reproduced, stored in a retrieval system, or transmitted
in any form or by any means, electronic, mechanical,

Longman Scientific & Technical
an imprint of
Longman Group Limited
Longman House, Burnt Mill, Harlow
Essex CM20 2JE, England
Associated companies throughout the world

First printed in Great Britain for Pitman Publishing Limited 1975
Reprinted (with corrections and amendments) 1978
Reprinted 1981, 1982, 1984
Reprinted by Longman Scientific & Technical 1986

ISBN 0-582-98855-1

Produced by Longman Singapore Publishers (Pte) Ltd.
Printed in Singapore.

Preface

This book contains what we consider to be the "basics" of
Engineering Drawing. Orthographic Projection, Conventions,
Sectioning, Pictorial Representation and Dimensioning have
been covered in detail as we feel that a thorough understanding of these topics forms a sound foundation upon which
to build. All technical information, examples, exercises
and solutions have been compiled in accordance with the
latest "metric" drawing office standards - B.S. 308:1972.

The book has not been written for any specific course
but can be profitably used both by students being introduced
to Engineering Drawing and also by those who have acquired a
little knowledge of the subject and wish to consolidate and
increase their understanding by working through carefully
graded exercises. It should prove useful for Craft, Tech-
nician (T.E.C.), O.N.D., C.S.E. and G.C.E. students and
also for those H.N.D. and Degree students with little
drawing experience. The book is seen primarily as a student
self-educator though no doubt many teachers will find it
useful as a reference source and/or exercise "bank".

Topics have been presented in a similar manner wherever
possible. Generally the opening page introduces the topic,
the next imparts the basic facts - visually rather than
verbally wherever possible. An illustrative example is
provided to aid understanding and this is followed by a
series of carefully graded exercises.

We are well aware of the dangers of presenting exercises which are known to contain errors. They have been
included because in our experience they are the common
misconceptions among students of engineering drawing. In
all cases the correct method and answers are given, some-
times immediately following the example, or in the solutions
at the end of the book.

It must be emphasized that this book not only trans-
mits information it is also a work-book. Do not be afraid
of drawing and writing on the pages! If maximum benefit is
to be derived from the book then the old maxim, "I do and I
understand" must be the students' guide.

We thank those people whose observations and suggestions
have helped us improve upon the first edition of the book.
We also wish to thank the British Standards Institution for
allowing us to use extracts from B.S. 308:1972.

R.S.R. & L.B.C.

Contents

SOLUTIONS

Orthographic Projection

Communication

There are many different ways of communicating ideas, information, instructions, requests, etc. They can be transmitted by signs or gestures, by word of mouth, in writing, or graphically. In an industrial context the graphical method is commonly used, communication being achieved by means of engineering drawings.

If oral and written communication only were used when dealing with technical matters, misunderstandings could arise, particularly in relation to shape and size. The lack of a universal spoken language makes communication and understanding even more difficult because of the necessity to translate both words and meaning from one language to another.

However, the universally accepted methods used in graphical communication through engineering drawings eliminate many of these difficulties and make it possible for drawings prepared by a British designer to be correctly interpreted or "read" by, for example, his German, French or Dutch counterpart.

Equally important, the components shown on the drawings could be made by suitably skilled craftsmen of any nationality provided they can "read" an engineering drawing.

Conventionally prepared engineering drawings provide the main means of communication between the "ideas" men (the designers and draughtsmen) and the craftsmen (machinists, fitters, assemblers, etc.). For the communication to be effective, everyone concerned must interpret the drawings in the same way. Only then will the finished product be exactly as the designer envisaged it.

To ensure uniformity of interpretation the British Standards Institution have prepared a booklet entitled BS 308:1972, *Engineering Drawing Practice*. Now in three parts, this publication recommends the methods which should be adopted for the preparation of drawings used in the engineering industry.

The standards and conventions in most common use and hence those required for a basic understanding of Engineering Drawing are illustrated and explained in this book.

Orthographic Projection

In the engineering industry communication between the drawing office and the workshop is achieved mainly by means of engineering drawings. The principal method used to prepare these drawings is known as Orthographic Projection.

Basically, Orthographic Projection is the representation of a three-dimensional component on a flat surface (the drawing sheet) in two-dimensional form. At least two orthographic views, therefore, are required to indicate fully the shape and size of a component. If the component is a complicated one then usually more than two views are shown to aid understanding.

In this country two methods of Orthographic Projection are used. One is known as First Angle Orthographic Projection (often referred to as English Projection), the other as Third Angle Orthographic Projection (American Projection). Both methods of representation are illustrated and explained in this section.

Orthographic Projection: First Angle

The pictorial drawing opposite indicates the shape of the component with a single view.

An orthographic drawing indicates the shape of a component by using a number of views each looking at a different face of the component.

At least two views are necessary to fully represent the component. Usually, however, three views are shown in order to clarify internal and external detail:

(1) A Front View

(2) A Plan View

(3) A Side View

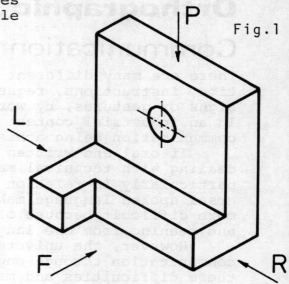
Fig.1

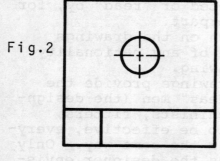
Fig.2

Front View (F)

(1) The front view, or front elevation, represents what is seen when looking at the front of the component in the direction of arrow F.

(2) The plan view represents what is seen when looking at the top of the component in the direction of arrow P at 90° to arrow F.

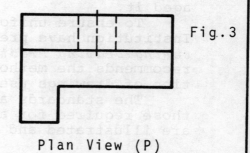
Fig.3

Plan View (P)

(3) A side view, or side elevation, represents what is seen when looking at the side of the component in the direction of either arrow R or arrow L. These arrows are at 90° to both arrow F and arrow P.

View looking in direction of arrow R.

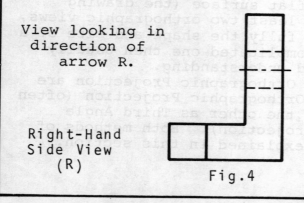

Right-Hand Side View (R)

Fig.4

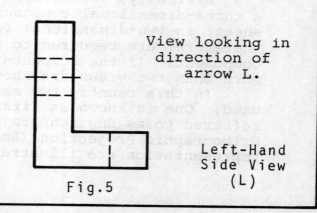

View looking in direction of arrow L.

Left-Hand Side View (L)

Fig.5

The separate views of the component are combined to form a complete orthographic drawing as shown below.

The front and side views are drawn in line with each other so that the side view may be "projected" from the front view and vice versa.

The plan view is drawn in line with and below the front view. In other words, the plan is projected from the front view.

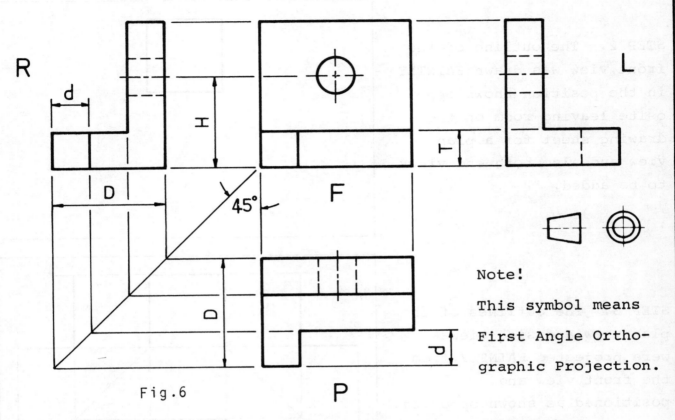

Fig.6

Note!

This symbol means First Angle Ortho-graphic Projection.

Points to note when making a drawing using First Angle Orthographic Projection:

(1) Corresponding heights in the front view and side view are the same.
    For example, the height of the hole from the base, H, is the same in both front and side views.
    The thickness of the base, T, is the same in both front and side views.

(2) Widths in the side view correspond to depths in the plan.
    For example, the total width, D, in the side view is the same as the total depth, D, in the plan.
    The width, d, is the same in both plan and side views.

    Projection of widths from side view to plan is made easier by using the 45° swing line as shown.

(3) The plan view is usually projected BELOW the front view.
    It can be above but this would be called an "inverted" plan.

(4) The R.H. side view is shown on the L.H. side of the front view.
    The L.H. side view is shown on the R.H. side of the front view.

    Note: Drawings should be read (or interpreted) by viewing from the R.H. side or bottom R.H. corner of the drawing.

The orthographic drawing of the bracket, Fig.6, was constructed step by step as follows:

STEP 1. The face to be used as the front view of the component was chosen, in this case, looking in the direction of arrow F (Fig.1). The selection of the front view is purely arbitrary.

STEP 2. The outline of the front view was drawn FAINTLY in the position shown opposite leaving room on the drawing sheet for a plan view and also both end views to be added.

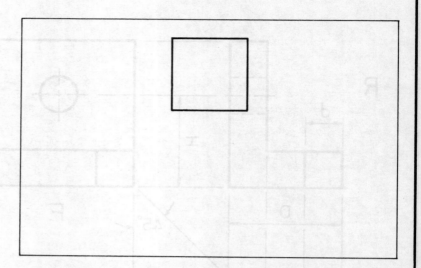

STEP 3. The outlines of the plan view and side views were projected FAINTLY from the front view and positioned as shown opposite.

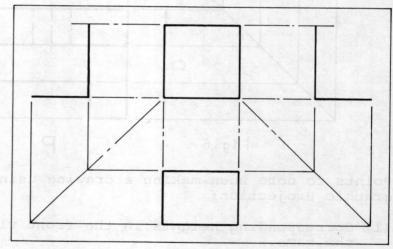

STEP 4. All remaining external details were added and centre lines inserted as shown opposite.

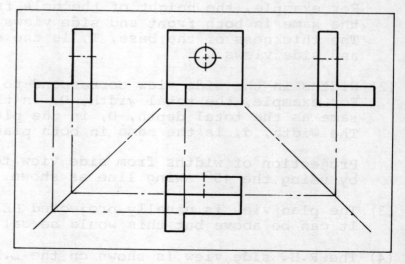

STEP 5. All hidden detail, i.e. for hole and recess, was added and the outline "heavied-in" to complete the drawing as shown in Fig.6.

Lines

For general engineering drawings, the following types of lines should be used.

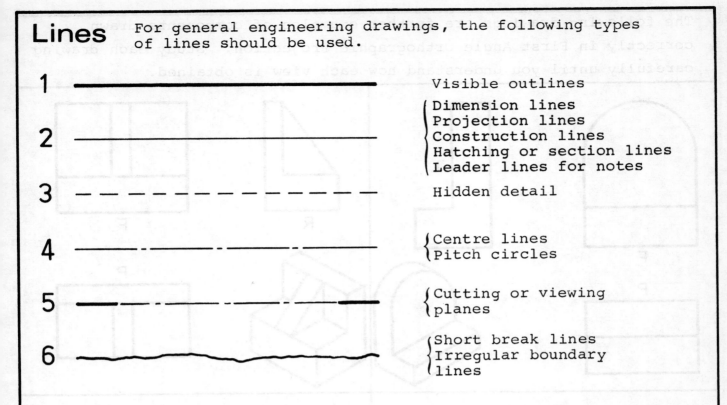

1 ————————————————————	Visible outlines
2 ————————————————————	{ Dimension lines Projection lines Construction lines Hatching or section lines Leader lines for notes
3 – – – – – – – – – – –	Hidden detail
4 ——— - ——— - ——— - ———	{ Centre lines Pitch circles
5 ▬▬ - ▬▬ - ▬▬ - ▬▬	{ Cutting or viewing planes
6 ∿∿∿∿∿∿∿∿∿∿	{ Short break lines Irregular boundary lines

Typical applications of some of the recommended types of lines have been shown in previous figures and are further illustrated below.

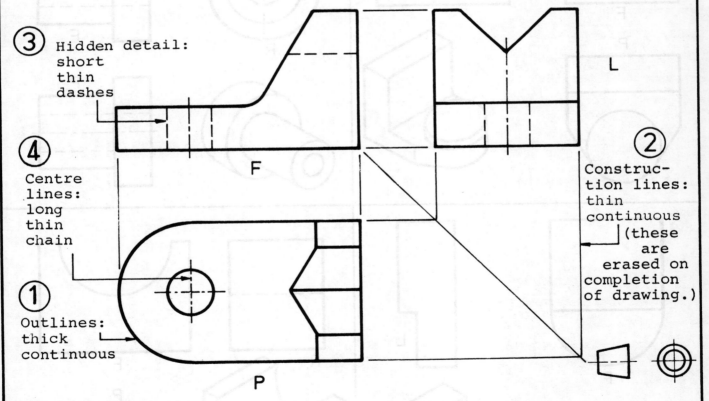

③ Hidden detail: short thin dashes

L

④ Centre lines: long thin chain

② Construction lines: thin continuous (these are erased on completion of drawing.)

① Outlines: thick continuous

F

P

HIDDEN DETAIL

The line numbered 3 above is used to represent hidden detail, i.e. edges, holes, surfaces, etc., which are known to exist but cannot be seen when viewed from outside the component.

Note: A hidden detail line is a broken line *not* a dotted line

5

The following drawings are further examples of components drawn correctly in First Angle Orthographic Projection. Study each drawing carefully until you understand how each view is obtained.

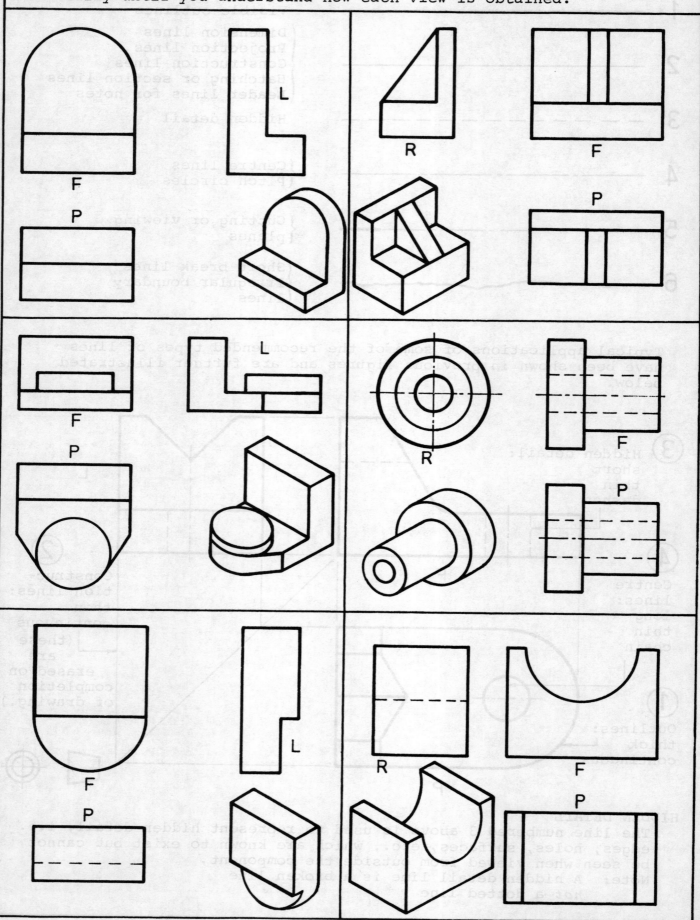

Drawings 1 to 9, shown in First Angle Orthographic Projection, represent the component below.

Examine each drawing carefully and explain briefly, either in the table below or on a separate sheet of paper, why each representation is INCORRECT. Solns. p.137

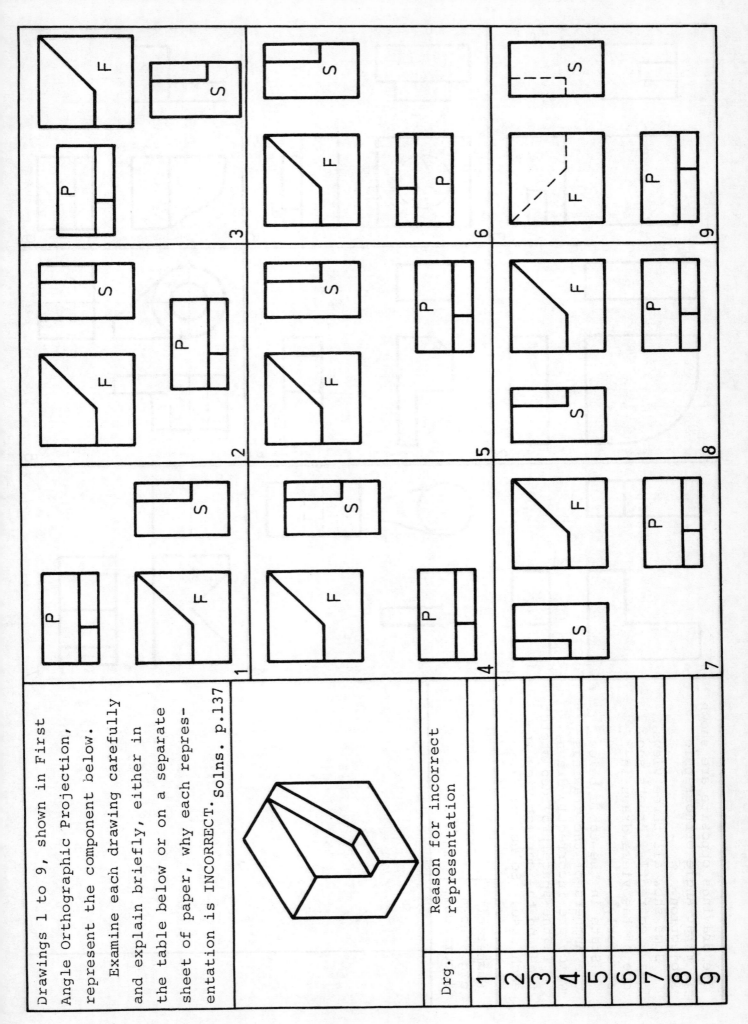

Drg.	Reason for incorrect representation
1	
2	
3	
4	
5	
6	
7	
8	
9	

The drawings opposite are shown in First Angle Orthographic Projection.

Some show all three views drawn correctly, others show one of the views drawn incorrectly.

State in the table below the numbers of the drawings which are incorrectly drawn.

In the space provided sketch freehand the correct view in each case. Solns. p.137

Number of drg. shown incorrectly	Sketch of correct view

8

Select, from the views A to L below, the missing view from each of the drawings 1 to 12. Insert the letter in the space provided.
Example: the missing view from drawing number 1 is C. **Solns. p.137**

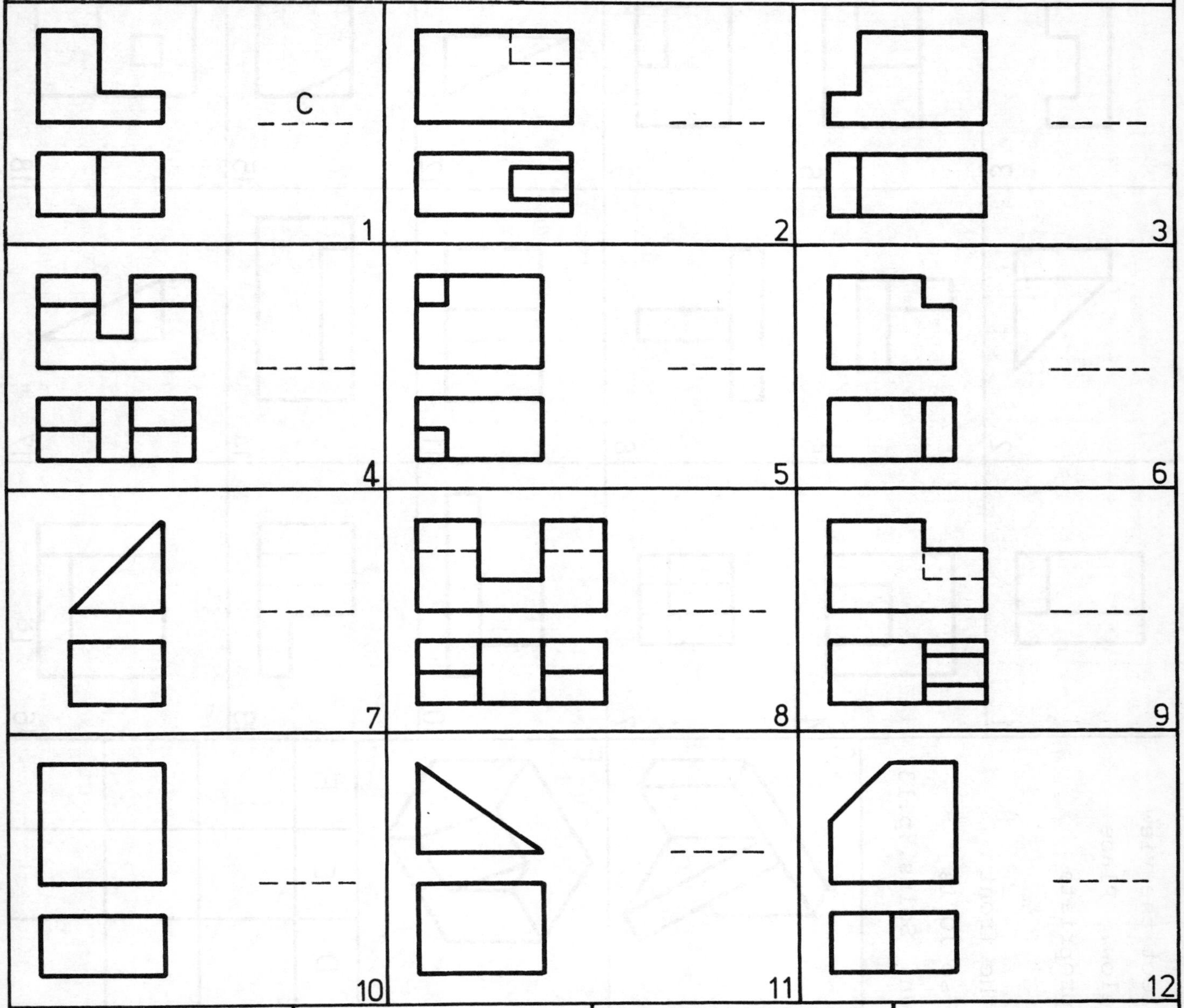

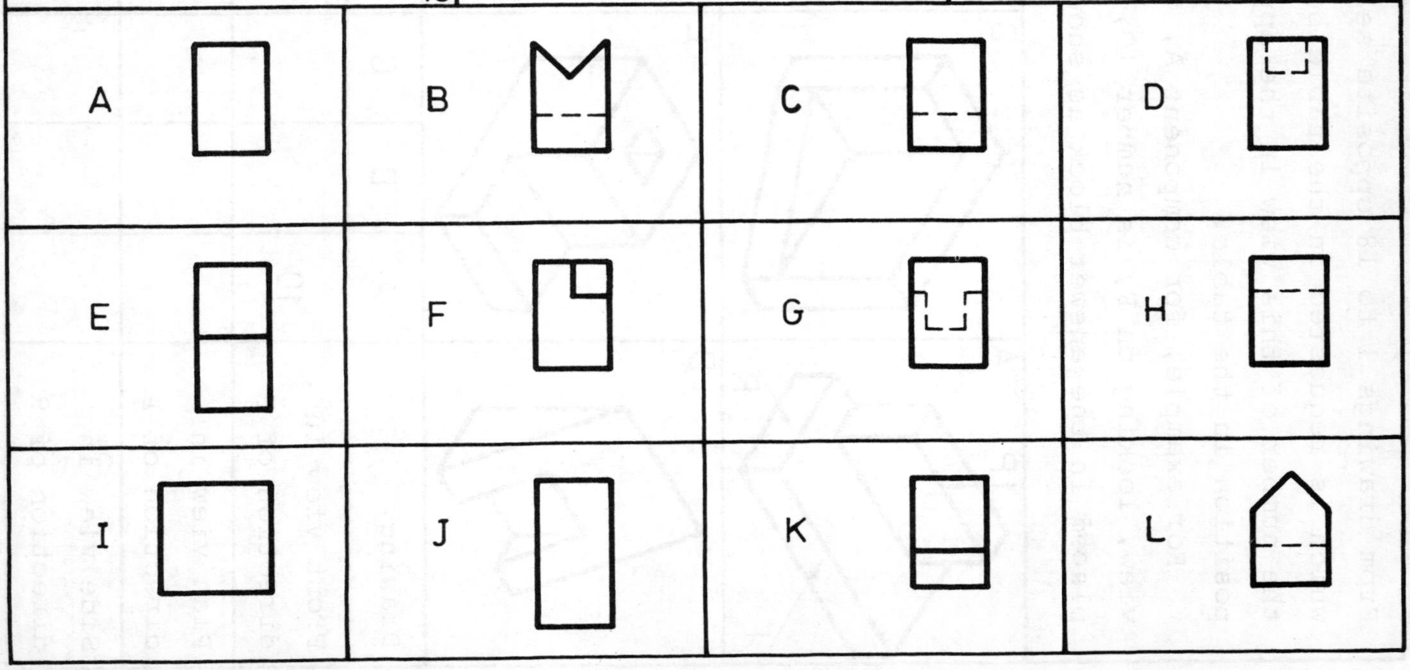

1 _ _ C _

2 _ _ _ _

3 _ _ _ _

4 _ _ _ _

5 _ _ _ _

6 _ _ _ _

7 _ _ _ _

8 _ _ _ _

9 _ _ _ _

10 _ _ _ _

11 _ _ _ _

12 _ _ _ _

A

B

C

D

E

F

G

H

I

J

K

L

9

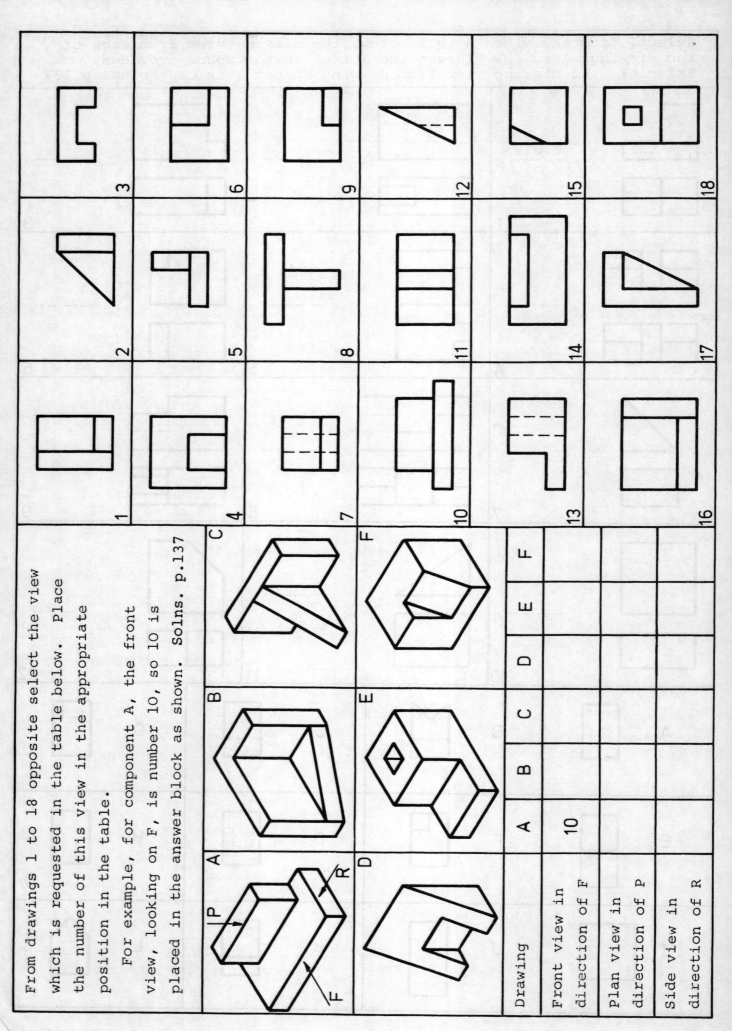

From drawings 1 to 18 opposite select the view which is requested in the table below. Place the number of this view in the appropriate position in the table.

For example, for component A, the front view, looking on F, is number 10, so 10 is placed in the answer block as shown. Solns. p.137

Drawing	A	B	C	D	E	F
Front view in direction of F	10					
Plan view in direction of P						
Side view in direction of R						

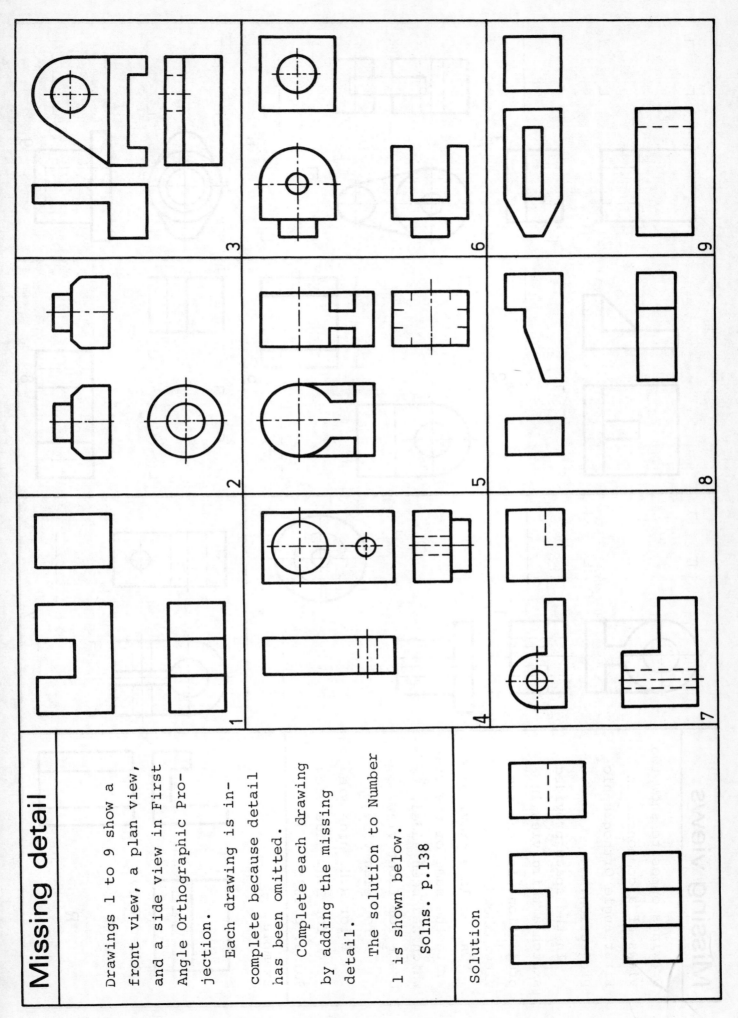

Missing detail

Drawings 1 to 9 show a front view, a plan view, and a side view in First Angle Orthographic Projection.

Each drawing is incomplete because detail has been omitted.

Complete each drawing by adding the missing detail.

The solution to Number 1 is shown below.

Solns. p.138

Solution

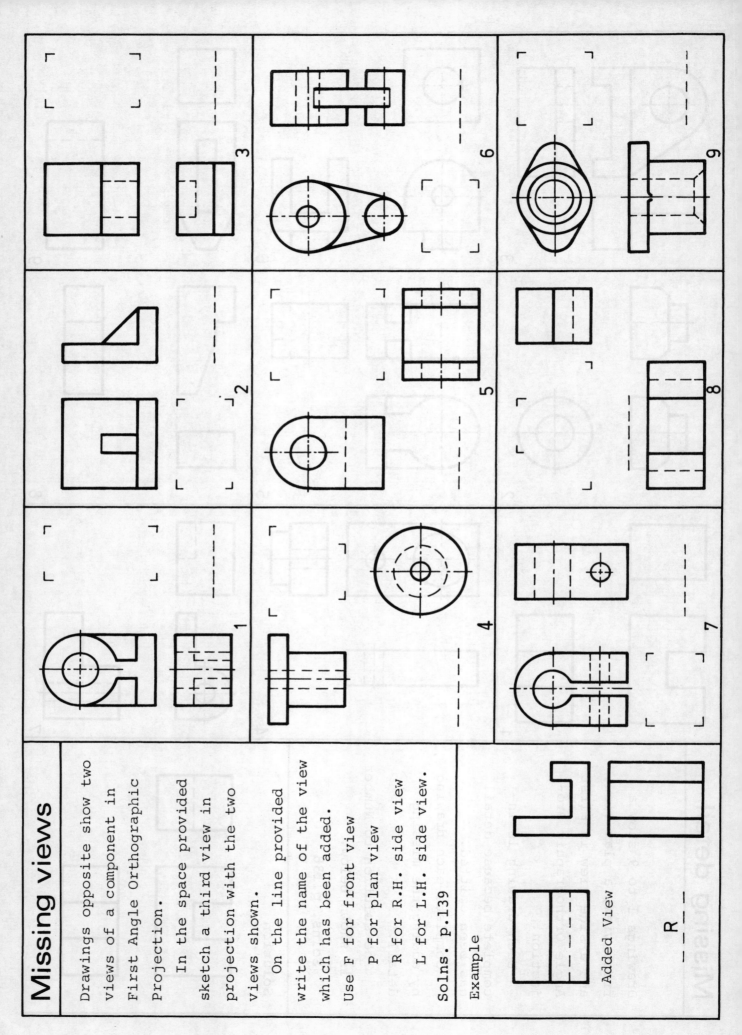

Missing views

Drawings opposite show two
views of a component in
First Angle Orthographic
Projection.

In the space provided
sketch a third view in
projection with the two
views shown.

On the line provided
write the name of the view
which has been added.

Use F for front view
P for plan view
R for R.H. side view
L for L.H. side view.

Solns. p.139

Example

R

Added View

Sketching orthographic views

The components shown below are drawn pictorially.

For each component sketch in good proportion in the space provided, a front view, a plan view and a l.h. side view.
Sketch the front view looking from F.
Show all hidden detail.
Use first angle projection as shown on the right.

Example
Solns. p.139

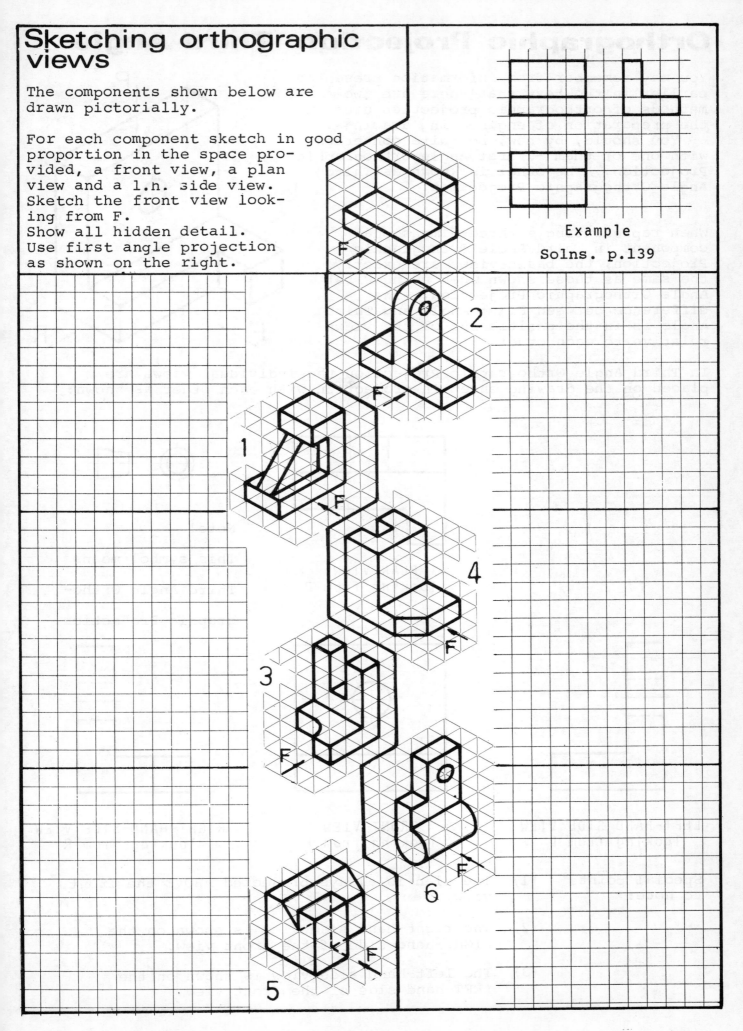

Orthographic Projection: Third Angle

You may remember from information presented earlier in the text that there are two methods of orthographic projection used in the preparation of engineering drawings.

You should, by now, be quite conversant with one of them - First Angle Orthographic Projection. The other is known as Third Angle Orthographic Projection.

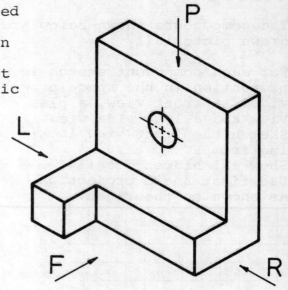

When representing a three-dimensional component in Third Angle Orthographic Projection, the basic views are exactly the same as those shown when using First Angle Orthographic Projection. The difference between First Angle and Third Angle is in the positioning of the views relative to each other.

In Third Angle Orthographic Projection the individual views are placed on the drawing sheet in projection with each other as shown:

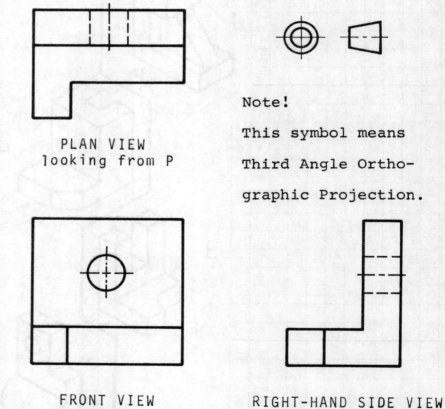

PLAN VIEW
looking from P

Note!

This symbol means Third Angle Ortho-graphic Projection.

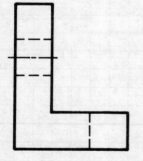

LEFT-HAND SIDE VIEW
looking from L

FRONT VIEW
looking from F

RIGHT-HAND SIDE VIEW
looking from R

Special points to note:

(1) The plan is always projected ABOVE the front view.

(2) The right-hand side view is shown on the RIGHT-hand side of the front view.

(3) The left-hand side view is shown on the LEFT-hand side of the front view.

A comparison of first angle projection and third angle projection

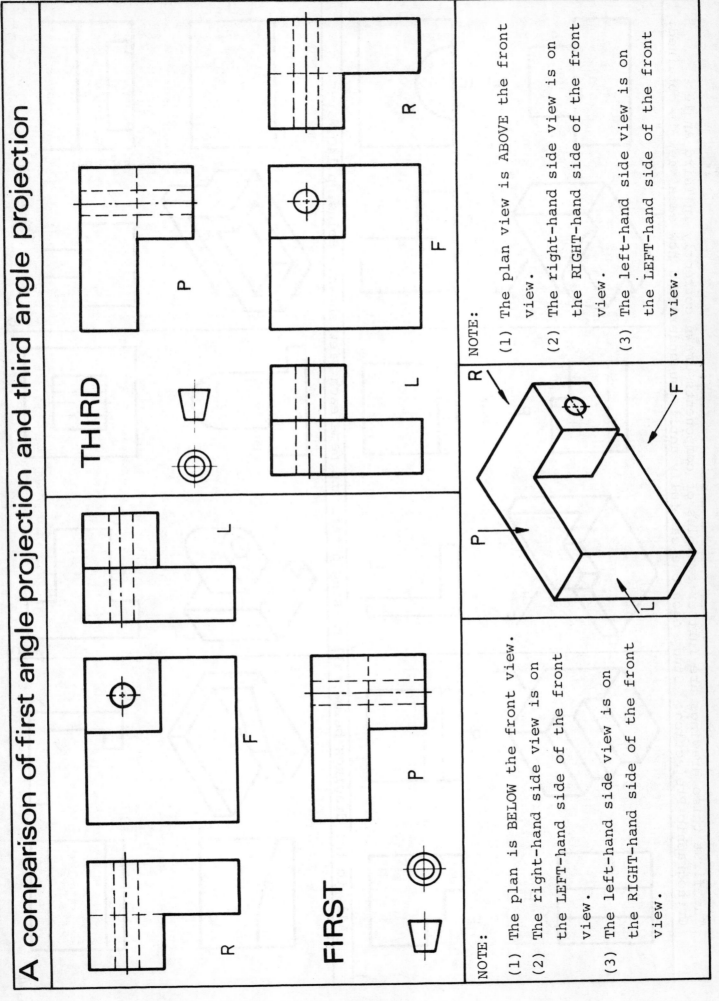

THIRD

P

R

F

L

FIRST

R

L

F

P

NOTE:

(1) The plan is BELOW the front view.

(2) The right-hand side view is on the LEFT-hand side of the front view.

(3) The left-hand side view is on the RIGHT-hand side of the front view.

NOTE:

(1) The plan view is ABOVE the front view.

(2) The right-hand side view is on the RIGHT-hand side of the front view.

(3) The left-hand side view is on the LEFT-hand side of the front view.

R

P

L

F

The first three drawings are further examples of components drawn correctly in Third Angle Orthographic Projection. Study each one carefully until you understand how each view is obtained. Solns. p.140.

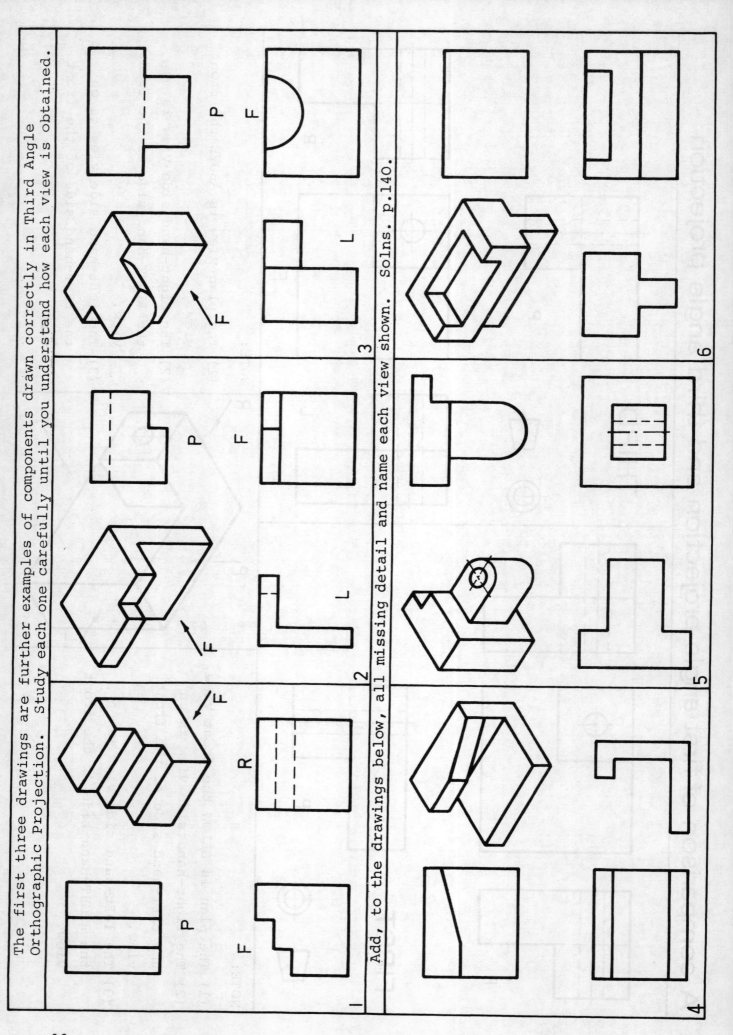

Add, to the drawings below, all missing detail and name each view shown.

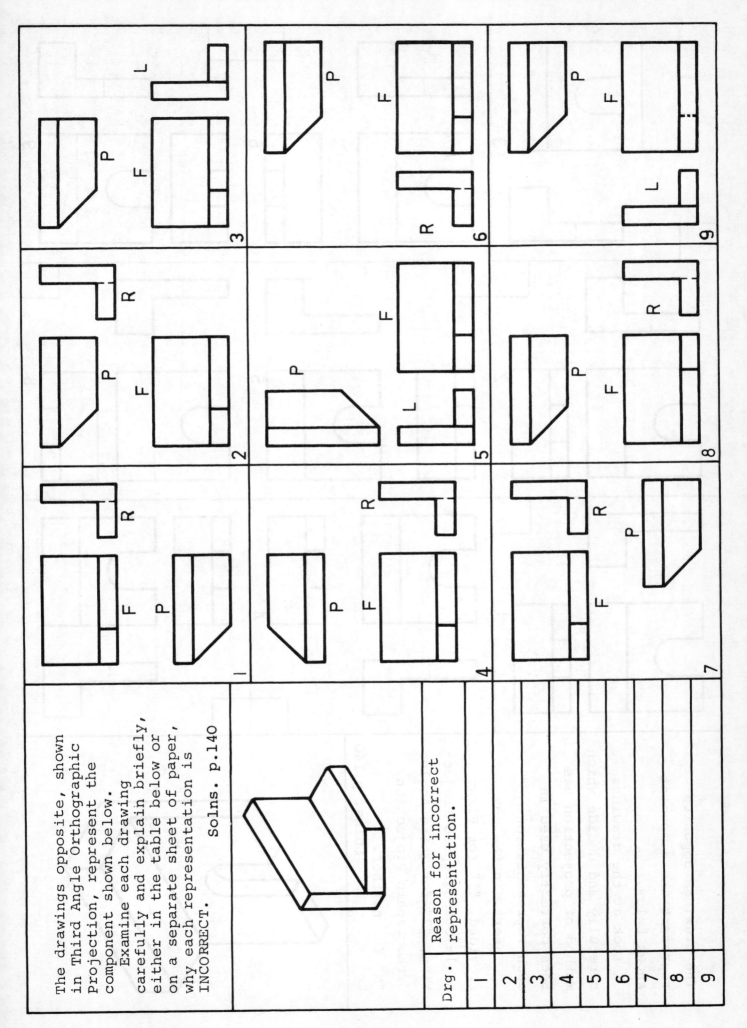

The drawings opposite, shown in Third Angle Orthographic Projection, represent the component shown below.

Examine each drawing carefully and explain briefly, either in the table below or on a separate sheet of paper, why each representation is INCORRECT.

Solns. p.140

Drg.	Reason for incorrect representation.
1	
2	
3	
4	
5	
6	
7	
8	
9	

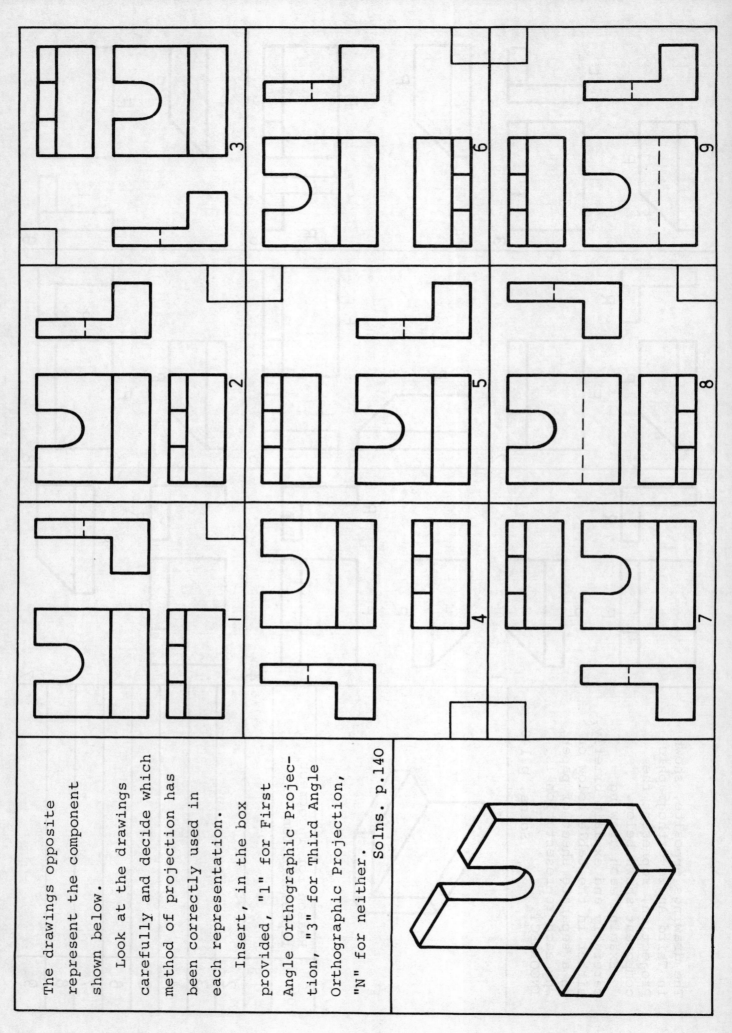

The drawings opposite represent the component shown below.

Look at the drawings carefully and decide which method of projection has been correctly used in each representation.

Insert, in the box provided, "1" for First Angle Orthographic Projection, "3" for Third Angle Orthographic Projection, "N" for neither.

Solns. p.140

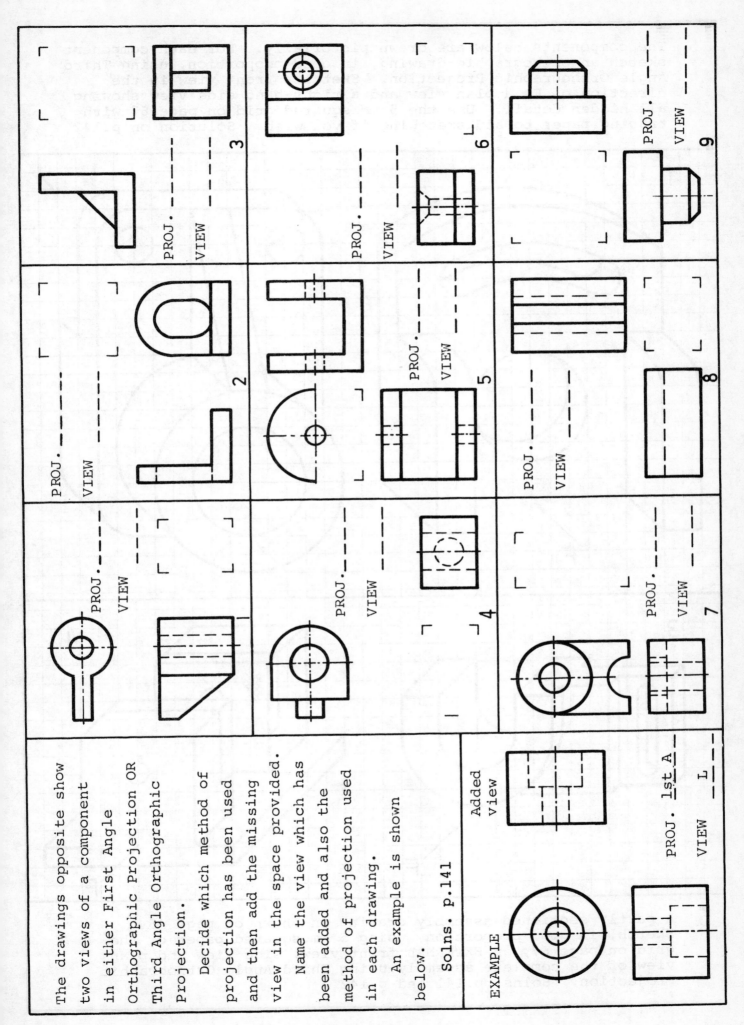

The drawings opposite show two views of a component in either First Angle Orthographic Projection OR Third Angle Orthographic Projection.

Decide which method of projection has been used and then add the missing view in the space provided. Name the view which has been added and also the method of projection used in each drawing.

An example is shown below.

Solns. p.141

EXAMPLE

PROJ. 1st A

VIEW L

Added view

1

2

3

4

5

6

7

8

9

PROJ.

VIEW

The components below are drawn pictorially. For each component sketch an orthographic drawing, in good proportion, using Third Angle Orthographic Projection. Sketch a front view in the direction of F, a plan view and a right-hand side view showing all hidden detail. Use the 5 mm squared grid on page 67 with tracing paper to aid sketching if you wish. Solution on p.142.

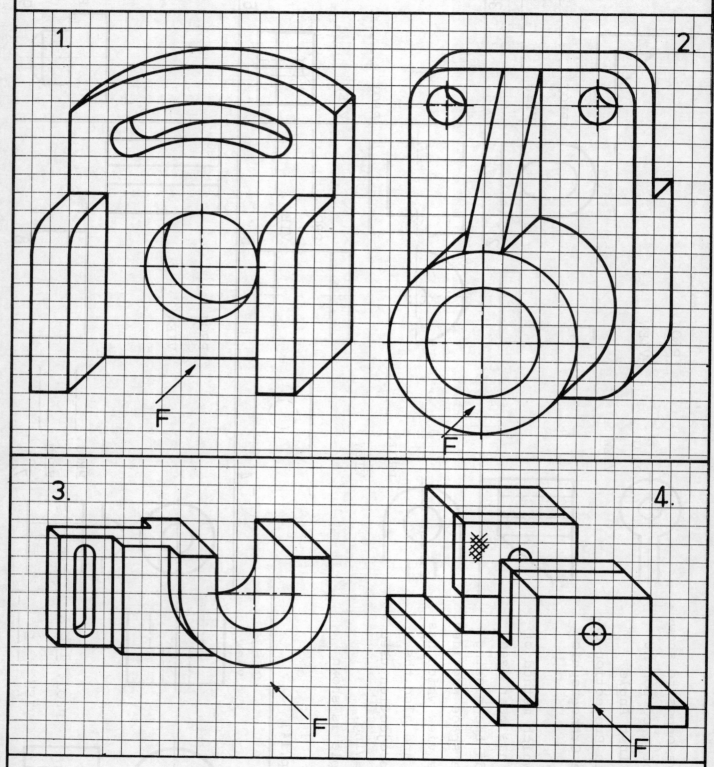

5. A partly sectioned assembly drawing in shown on page 48. Sketch, in good proportion, using 5 mm squared paper or the grid on page 67, an EXTERNAL front view, side view and plan view of the complete assembly using Third Angle Orthographic Projection. Solns. p.141 and p.142

Sectioning

Drawings of the outside of simple components are often sufficient to convey all the information necessary to make the component. More complicated components, however, may require sectional views to clarify internal details.

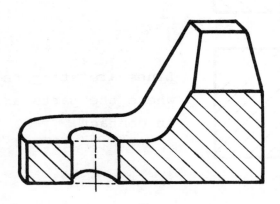

A sectional view is obtained when one imagines the component to be cut through a chosen section plane - often on a centre line.

If the vee-block is cut on section plane C-C as shown above, the resulting sectional view projected from the plan replaces the usual front view of the outside of the block.

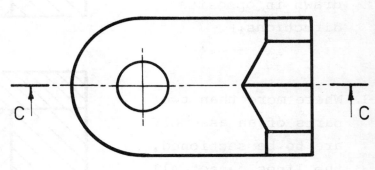

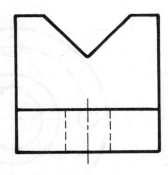

End view

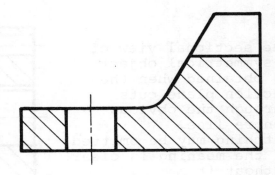

Sectional Front View
looking on cutting plane C-C

Sectional views are drawn only when it is necessary to explain the construction of a complex object or assembly. Some of the examples used in the next few pages have been chosen to illustrate the rules of sectioning although in practice, as in the case of the vee-block drawn above, a sectional view may not have been necessary

The draughtsman has to decide how a component or assembly should be sectioned in order to provide the fullest possible information. The recommendations of BS 308 enable him to do this in a way that is understood by all engineers.

The rules of sectioning:

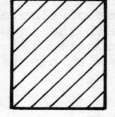

1. A sectioned object is shown by lines drawn preferably at 45°. Thin lines touch the outline. The lines are equi-spaced.

Size of sectioned part determines line spacing - preferably not less than 4 mm.

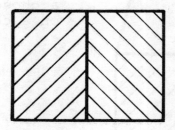

2. If two adjacent parts are sectioned, the section lines are drawn in opposite directions.

Lines are staggered where the parts are in contact.

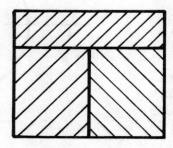

3. Where more than two parts of an assembly are to be sectioned, the lines cannot all be opposite.

Section lines are closer together on the third area — usually the smallest

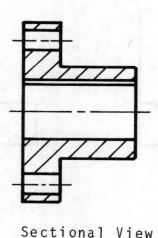

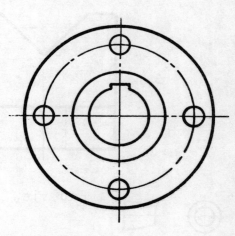

4. The sectional view of a symmetrical object is obtained when the section plane cuts through the obvious centre line. Hatching may be omitted if the meaning is clear without it.

Sectional View

5. If an object is NOT symmetrical the section plane chosen should be clearly stated.

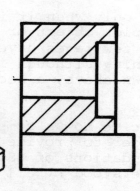

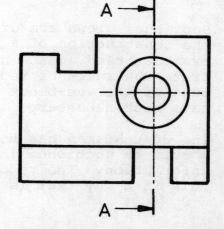

Section A-A

Sectioning: exceptions

There are a number of features and parts which are NOT normally
sectioned even though they may lie in the section plane. A
good way to accept these exceptions to the general rule is to
imagine how complicated the drawing would look if they were
sectioned. They are sectioned, however, when they lie ACROSS
the section plane. See section D-D. ✳

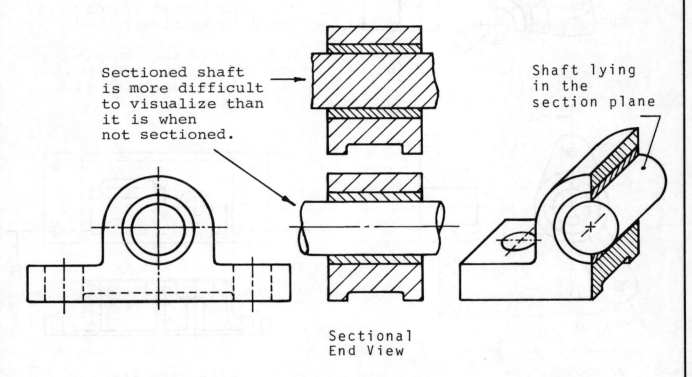

Sectioned shaft
is more difficult
to visualize than
it is when
not sectioned.

Shaft lying
in the
section plane

Sectional
End View

BS 308 states that the following are NOT sectioned even if they
lie in a given section plane:

| Shafts | Keys | Rivets | Ribs | Pins | Dowels |
| Cotters | Bolts | Gear Wheels | Nuts | Webs | Washers |

Some of these are illustrated in BS 308 by this diagram:

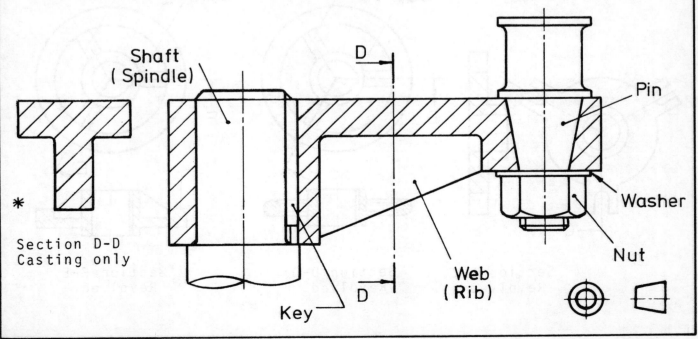

Shaft
(Spindle)

D

Pin

✳

Section D-D
Casting only

Washer

Nut

Web
(Rib)

Key

D

Staggered section planes

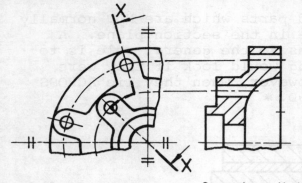

Examples

Section X-X

Each part of the section plane is swung to the vertical before projecting to the sectional End View.

By using this convention the draughtsman avoids using too many auxiliary views.

A staggered section plane should only be used when there is a resulting gain in clarity.

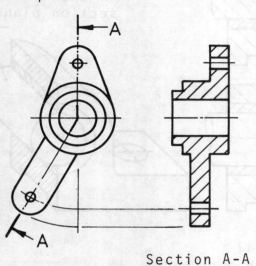

Section A-A
Re-aligned

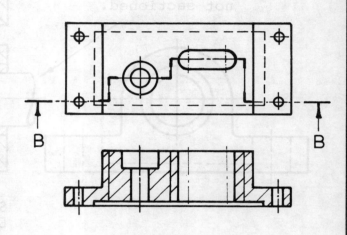

Section B-B
Staggered or offset

Applications of this convention to specific cases are given below.

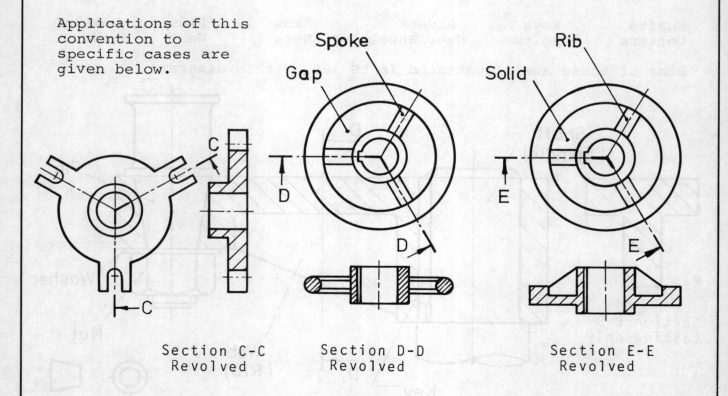

Spoke

Gap

Rib

Solid

Section C-C
Revolved

Section D-D
Revolved

Section E-E
Revolved

Errors in sectioning occur in each of the drawings below.
Trace or re-draw a correct sectional view in each case.
State the method of projection used where applicable.

Solns. p.144

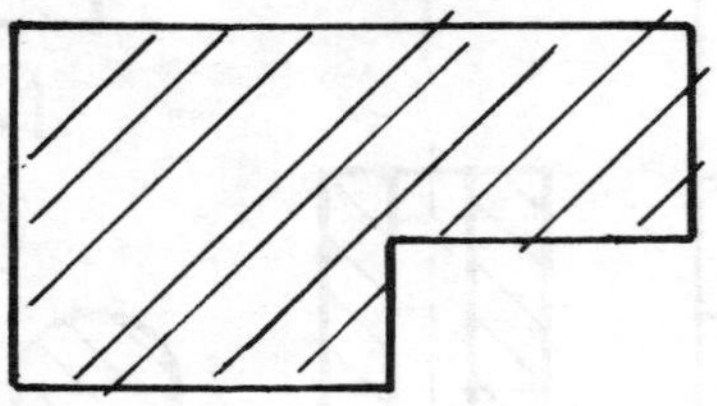

1

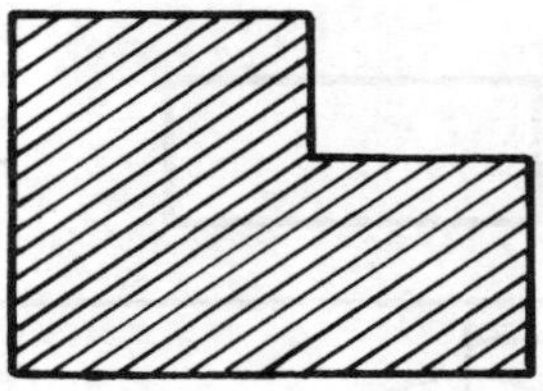

2

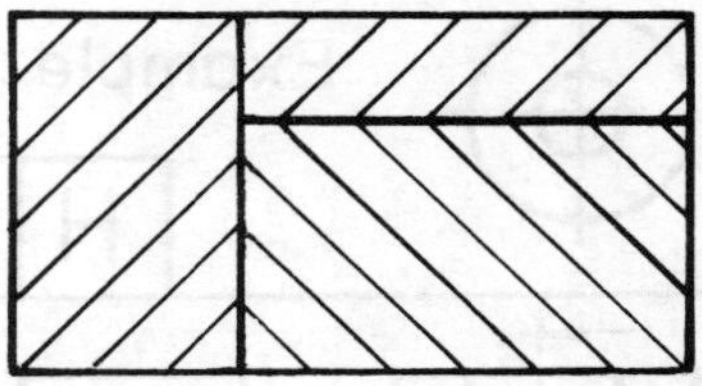

3

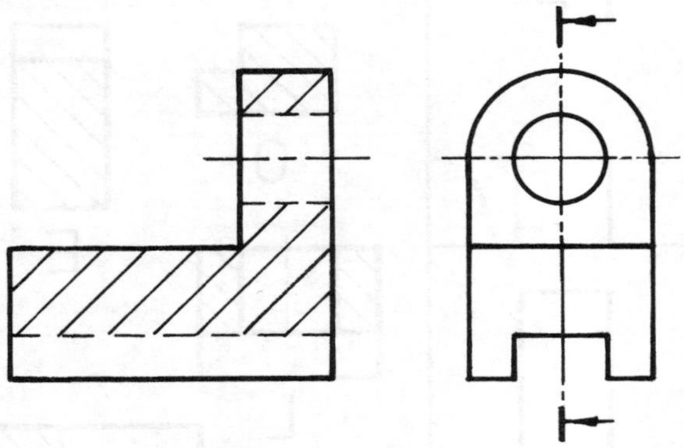

4

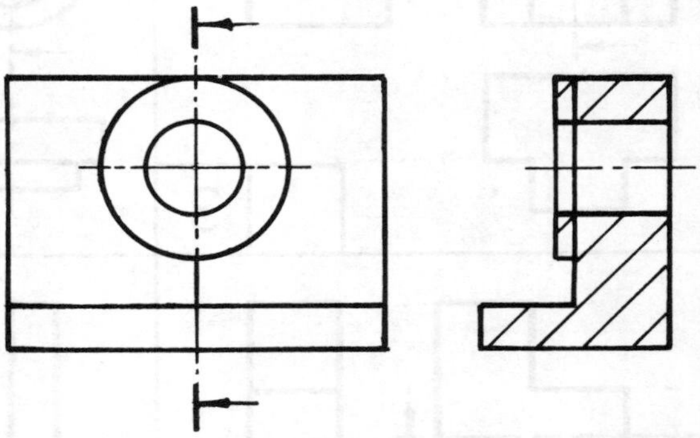

5

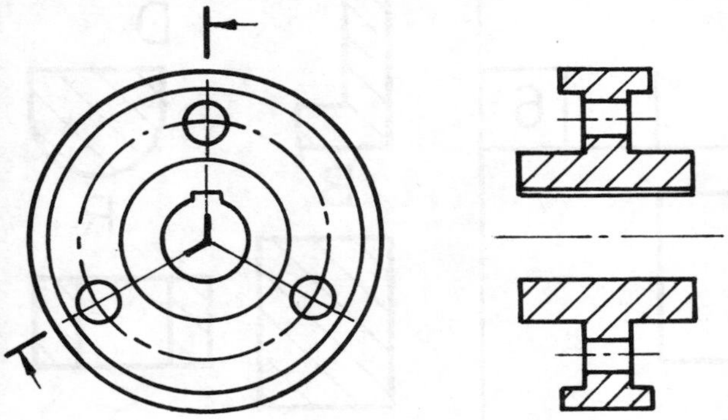

6

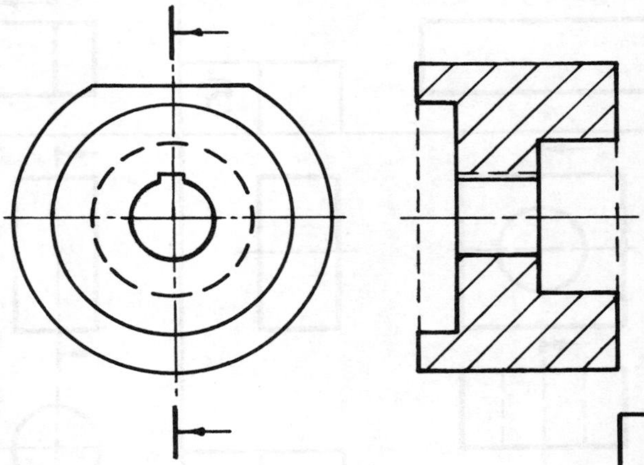

7

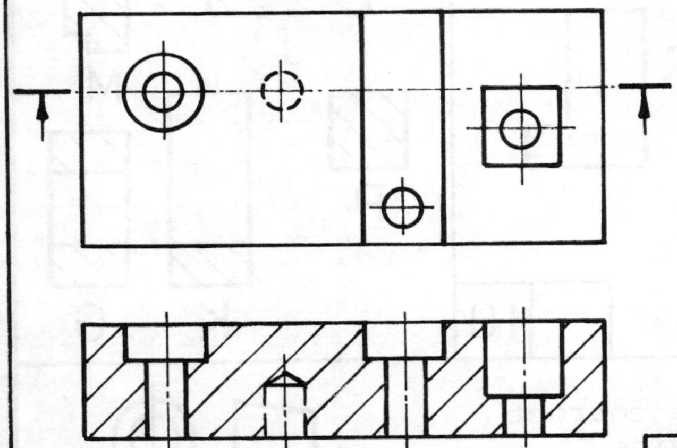

8

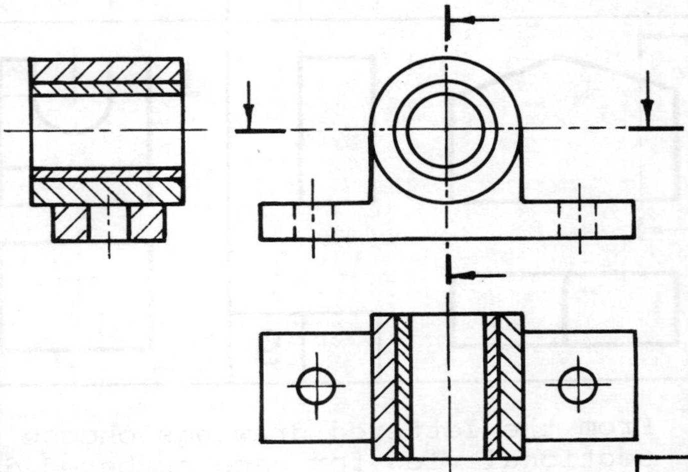

9

Example

H I

From the lettered drawings choose the correct sectional view for each numbered drawing. Sketch the view in the space provided. Solns. p.144

From the lettered drawings choose the correct
sectional view for each numbered drawing.
Sketch the view in the space provided. Solns. p.144

Sectioning exercises

Solns. p.145

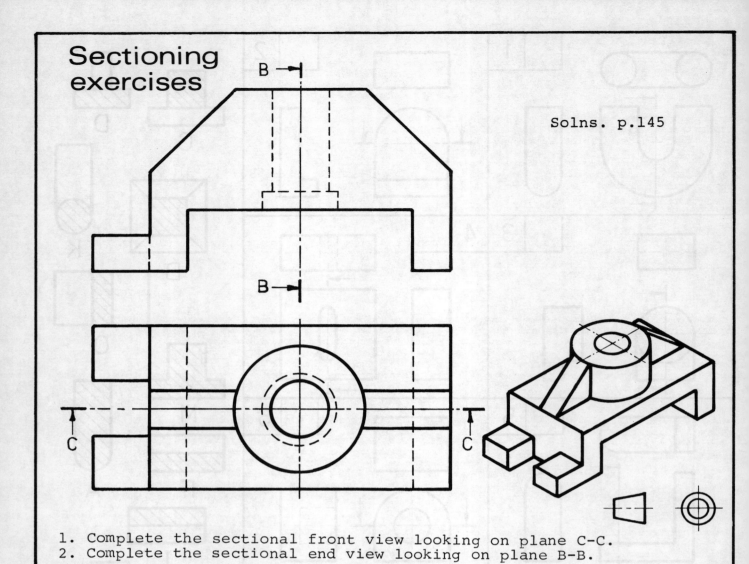

1. Complete the sectional front view looking on plane C-C.
2. Complete the sectional end view looking on plane B-B.

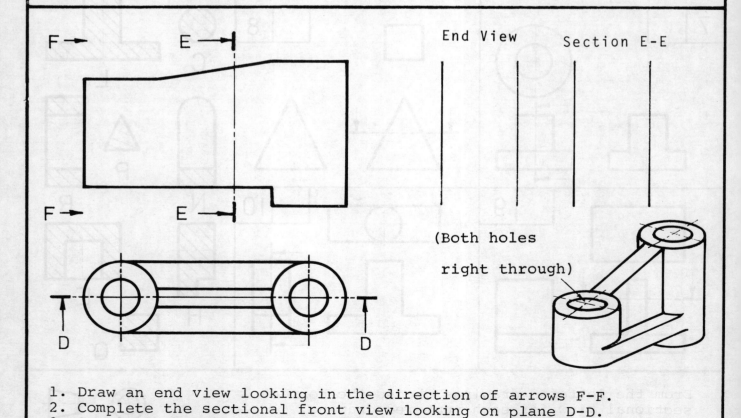

End View   Section E-E

(Both holes

right through)

1. Draw an end view looking in the direction of arrows F-F.
2. Complete the sectional front view looking on plane D-D.
3. Complete the sectional end view looking on plane E-E.

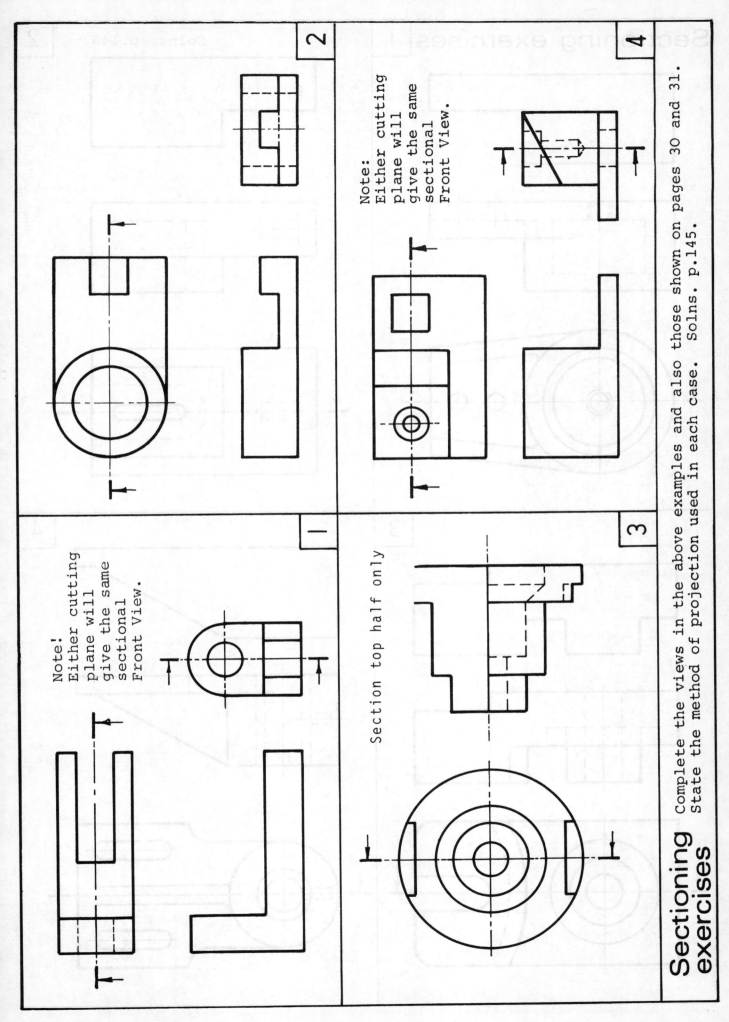

Sectioning exercises

Complete the views in the above examples and also those shown on pages 30 and 31. State the method of projection used in each case. Solns. p.145.

Note:
Either cutting plane will give the same sectional Front View.

Note:
Either cutting plane will give the same sectional Front View.

Section top half only

Note!
Either cutting plane will give the same sectional Front View.

2

4

1

3

Solns. p.146

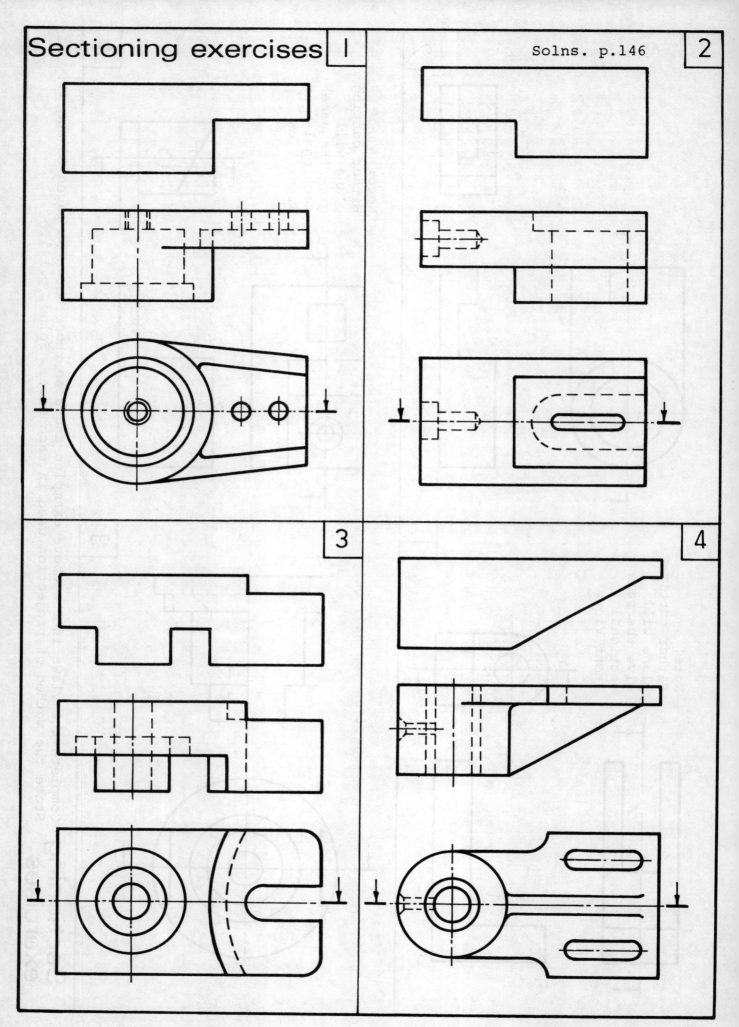

Solns. p.146

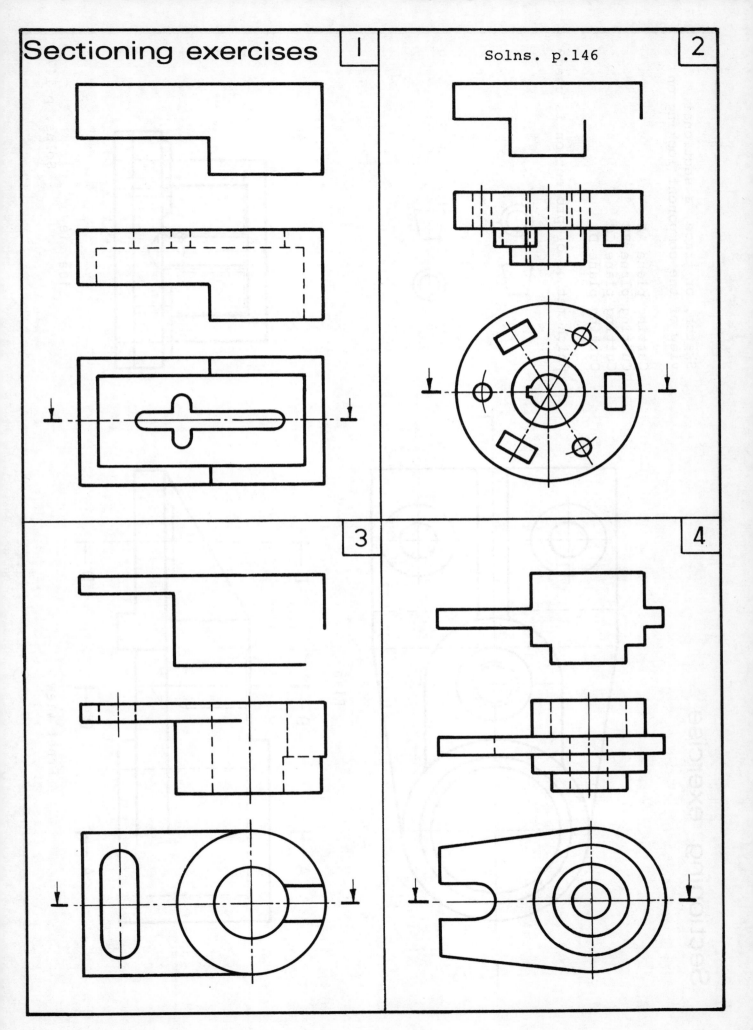

Sectioning exercise

Sketch, or trace, a sectional view of the component looking on

Cutting plane AA
Cutting plane BB
Cutting plane CC
Cutting plane DD

Which method of projection is used?

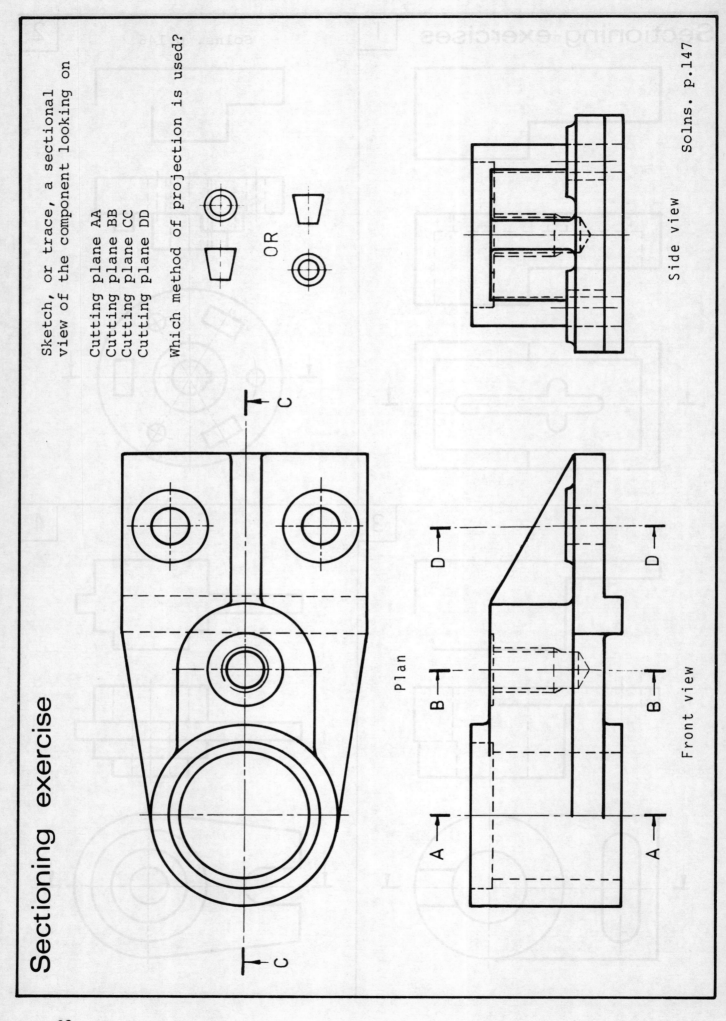

OR

Side view

Plan

Front view

Solns. p.147

Information on Engineering Drawings

Communication between the drawing office and the workshop is mainly achieved via the engineering drawing - orthographic and/or pictorial. In order to reduce drafting time a number of standard parts are drawn in a simplified form and many items of written information are abbreviated. In BS 308 Part 1:1972 recommendations have been made for

Conventional representation of commonly used parts & materials and Abbreviations of terms frequently used on drawings.

Before this engineers' "shorthand" can be used correctly it is necessary to understand the terms used to describe features of engineering components. This terminology is common to both drawing office and workshop and is often used when discussing the various manufacturing and machining processes used in engineering.

Terminology

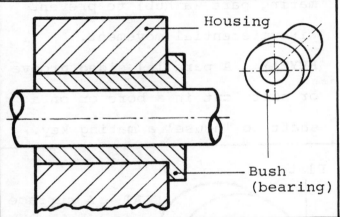

Housing

Bush (bearing)

HOUSING   A component into which a "male" mating part fits, sits or is "housed".

BUSH   A removable sleeve or liner. Known alternatively as a simple bearing. Fig. shows a flanged bush.

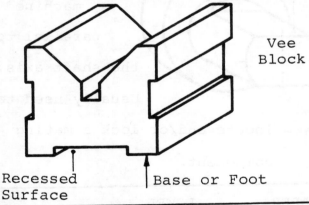

Vee Block

Recessed Surface

Base or Foot

BASE (or FOOT)   That part upon which the component rests.

RECESSED SURFACE   Ensures better seating of the base. Minimises machining of the base.

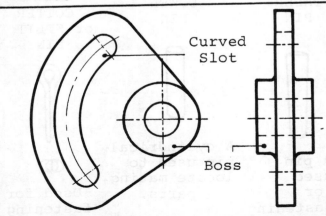

Curved Slot

Boss

BOSS   A cylindrical projection on the surface of a component - usually a casting or a forging.

CURVED SLOT   An elongated hole whose centre line lies on an arc. Usually used on components whose position has to be adjusted.

33

Terminology

Rib

Fillet

FILLET A rounded portion, or radius, suppressing a sharp internal corner.

RIB A reinforcement positioned to stiffen surfaces usually at right angles to each other

Key

Keyway

KEY A small block or wedge inserted between a shaft and mating part (a hub) to prevent circumferential movement.

KEYWAY A parallel sided groove or slot cut in a bore or on a shaft to 'house' a mating key.

Tee Groove (slot)

TEE SLOT Machined to "house" mating fixing bolts and prevent them turning

Flat

FLAT A surface machined parallel to the shaft axis. Usually used to locate and/or lock a mating component.

HEXAGON HEAD BOLT

HEXAGON HEAD SCREW

STUD

Note thread lengths

TAPER PIN

A pin used for fastening

DOWEL PIN

A cylindrical pin used to locate mating parts.

COTTER or SPLIT PIN

Used for fastening

Terminology

Many different types of holes may be seen on engineering drawings. The more common ones, associated with drilling, reaming and/or tapping, are shown below in Fig.1. The name and where appropriate the application of each is indicated.

Figures 2 and 3 indicate the different terms which may be seen on drawings that are associated particularly with lathe work.

Fig.1

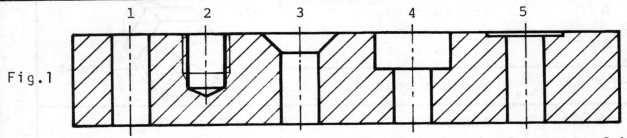

1. A *drilled hole* or, if greater accuracy is required, a reamed hole.
2. A *'blind' tapped hole* i.e. a threaded hole which passes only part way through the plate.
3. A *countersunk hole*. Provides a mating seat for a countersunk headed screw or rivet.
4. A *counterbore*. Provides a 'housing' for the heads of capscrews, bolts, etc.
5. A *spotface*. A much shallower circular recess. Provides a machined seat for nuts, bolt heads, washers, etc.

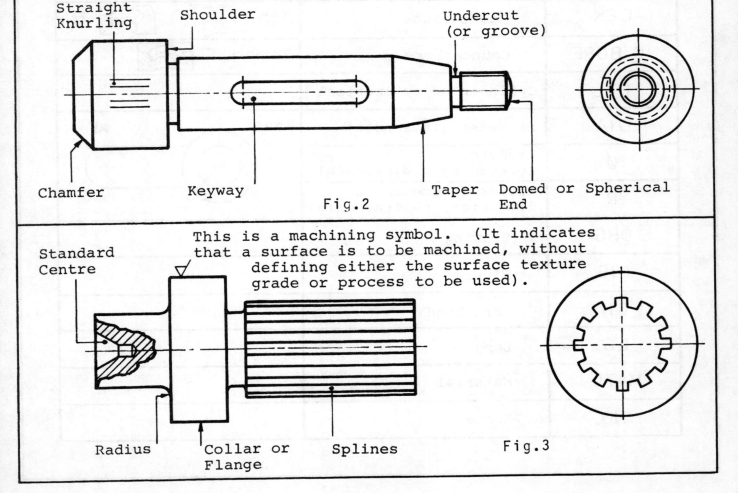

Straight Knurling   Shoulder   Undercut (or groove)

Chamfer   Keyway   Taper   Domed or Spherical End

Fig.2

Standard Centre   This is a machining symbol. (It indicates that a surface is to be machined, without defining either the surface texture grade or process to be used).

Radius   Collar or Flange   Splines   Fig.3

Abbreviations

Many terms and expressions in engineering need to be written on drawings so frequently as to justify the use of abbreviations which help to reduce drafting time and costs. Many of these abbreviations have been standardized as can be seen in BS 308:1972:section 11. A selection of the more commonly used ones are stated and clarified in the following tables.

ABBREVIATION	MEANING	SKETCH/NOTES
A/C	Across corners	
A/F	Across flats	
HEX HD	Hexagon head	
ASSY	Assembly	See page 96
CRS	Centres	25 CRS
℄ or CL	Centre line	
CHAM	Chamfered	CHAM
CH HD	Cheese head	CH HD SCREW
CSK	Countersunk	CSK SCREW CSK
C'BORE	Counterbore	C'BORE
CYL	Cylinder or cylindrical	
DIA	Diameter (in a note)	$\phi 20$ R5
∅	Diameter (preceding a dimension)	
R	Radius (preceding a dimension, capital only)	
DRG	Drawing	
FIG.	Figure	
LH	Left hand	
LG	Long	
MATL	Material	
NO.	Number	

ABBREVIATION	MEANING	SKETCH/NOTES
PATT NO.	Pattern number	
PCD	Pitch circle diameter	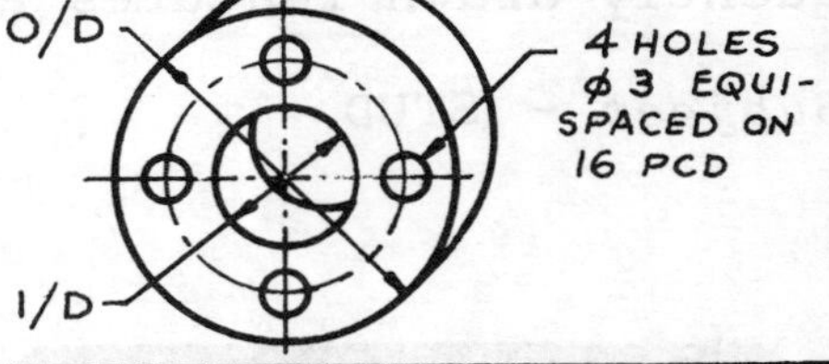
* I/D	Inside diameter	
* O/D	Outside diameter	O/D — 4 HOLES ⌀3 EQUI-SPACED ON 16 PCD / I/D
RH	Right hand	
RD HD	Round head	 RD HD SCREW
SCR	Screwed	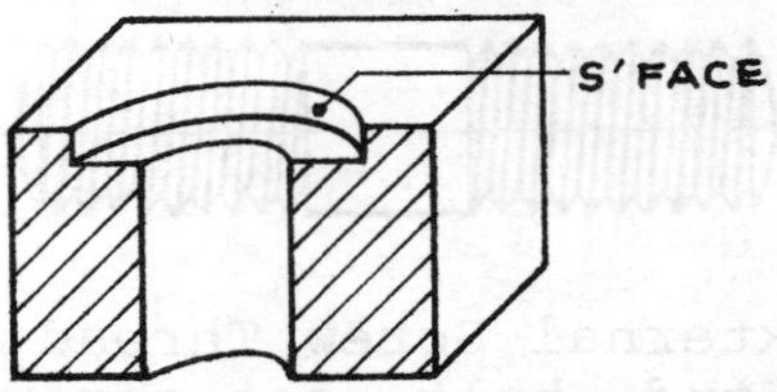 S'FACE
SPEC	Specification	
S'FACE	Spotface	
SQ	Square (in a note)	See page 40 for conventional representation
□	Square (preceding a dimension)	
STD	Standard	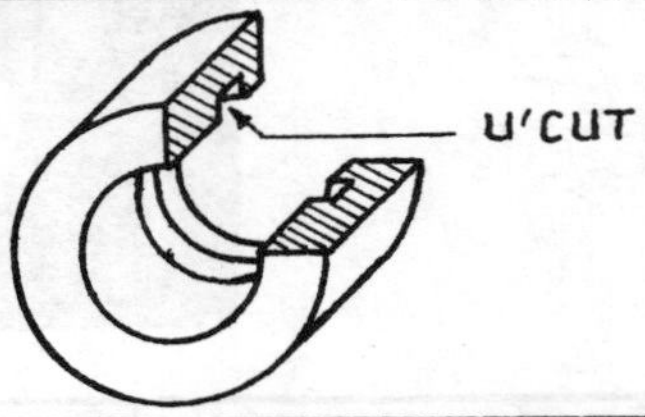
U'CUT	Undercut	U'CUT
* M/CD	Machined	
* mm	Millimetre	
* NTS	Not to scale	
* RPM	Revolutions per minute	SI symbol: rev/min
* SWG	Standard wire gauge	
* TPI	Threads per inch	

* Abbreviations commonly used in practice but not listed in

  BS 308: Part 1: 1972

Notes
  (1) Abbreviations are the same in the singular and plural.
  (2) Capital letters are shown above, lower case letters may
      be used where appropriate. For other abbreviations upper
      or lower case letters should be used as specified by
      other relevant British Standards.
  (3) Full stops are not used except when the abbreviation makes
      a word which may be confusing, e.g. NO. for 'number'.

Conventional Representation of Common Features: Screw threads

There are many components commonly used in engineering which are complicated to draw in full. In order to save drawing time, these parts are shown in a simplified, conventional form. Some of the more frequently drawn features are shown in this section.

Subject – STUD

Convention

The screw thread is represented by two parallel lines. The distance between these lines is approximately equal to the depth of thread.

Note   The inside line is THIN and the circle is *broken*

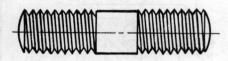

External Screw Thread
(Stud, bolt, set-screw, etc.)

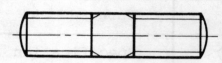

Front view   Side view

TAPPED
"BLIND HOLE"

The shape of the hole formed by the tapping size drill is drawn in heavy lines.

Note   Section lines *cross* the thread

The outside line is THIN and the circle is broken

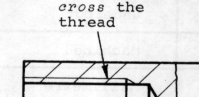

Internal Screw Thread

Front view   Side view

STUD INSIDE
TAPPED HOLE

The external thread is superimposed over the internal thread.

Note   Section lines are *not* drawn across the external thread.

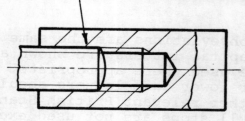

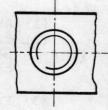

Screw Threads
(Assembly)

Front view   Side view

Conventional representation: Springs

A spring is designated by stating the diameter of the wire, the coil diameter (inside or outside), the form of the spring ends, the total number of coils and its free length - See BS 1726.

In the case of the compression spring, the pitch of the coils may be deduced from its free length and number of coils.

Subject　　　　　*Convention*

 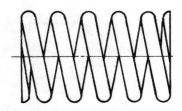 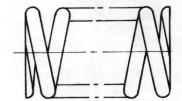

COMPRESSION SPRING

Note

The construction on the right shows how the convention may be drawn from the spring details. Extreme accuracy is not essential; for example, a radius gauge may be used for the small radii.

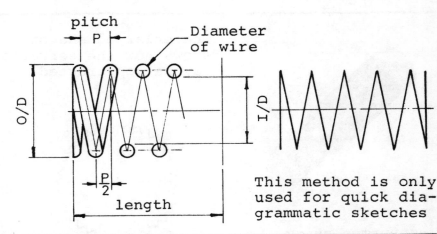

This method is only used for quick diagrammatic sketches

TENSION SPRING

Note

The construction on the right shows once again how the pitch of the coils may be used to set out the convention.

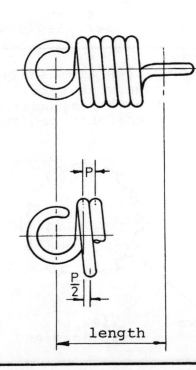 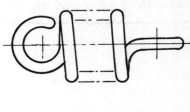

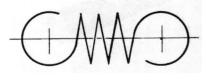

Diagrammatic representation

Conventional representation: Shaft Details

It is frequently necessary to fix a component to one end of a shaft or spindle so that a torque may be transmitted. Examples below are:

(1) Square on shaft - Machine handles, valve wheel spindle.
(2) Serrated shaft - Typical example - steering wheel to column on car.
(3) Splined shaft - Drive shaft on machine or vehicle when sliding also has to take place, e.g. in a gear box.

Subject   *Convention*

See page 96 for square on valve spindle.

SQUARE ON the end of a long SHAFT

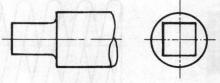

Side view

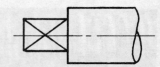

The enlarged sketch shows how the serrations may be constructed

SERRATED SHAFT

Note
Thirty-six serrations are shown for convenience only

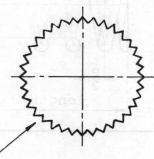

View on serrated end

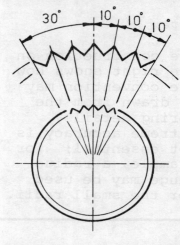

Note
Only a few teeth need be shown

Splines are usually drawn with parallel sides as emphasized in the enlarged sketch

SPLINED SHAFT

Note
Twelve splines are shown for convenience only

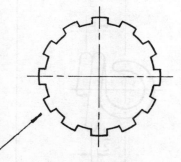

View on splined end

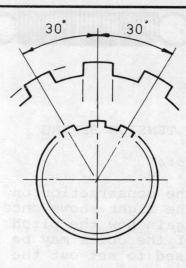

Refer to section on terminology - page 35

Conventional representation: Knurling

Knurling is a common method of providing a roughened surface to aid tightening or slackening of a screw by hand. This is formed by pressing special rollers against the surface of the component whilst it revolves in a lathe. The commonly used diamond and straight knurls are shown below.

Subject   *Convention*

 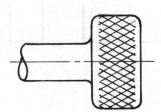 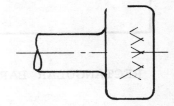

DIAMOND KNURL on
a machine screw head

 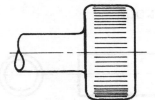 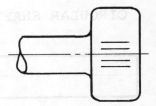

STRAIGHT KNURL on
a circuit terminal

Bearings

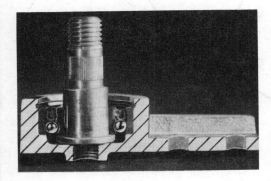

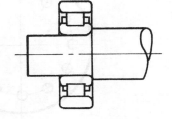

ROLLER

Notice how much less drawing is necessary in order to represent any ball or roller bearing in a conventional manner.

Cross-section of a double row self-aligning ball bearing.

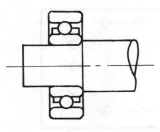 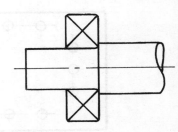

See assembly drawing on page 103.

BALL

Conventional representation: Long Components

There are occasions when bars, shafts, spindles or tubes may be too long to be drawn to a reasonable scale. In such cases the elevation may be interrupted as shown below.

Subject   *Convention*

RECTANGULAR BAR

CIRCULAR SHAFT OR SPINDLE

HOLLOW SHAFT OR TUBE

MULTIPLE HOLES

When a large number of holes of equal diameter are equi-spaced around a diameter or in line, only one hole need be drawn in full with the remainder marked with a short centre-line as shown in the drawings on the right.

Remember that this circle is called the pitch circle diameter or PCD

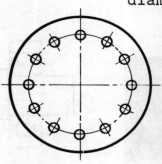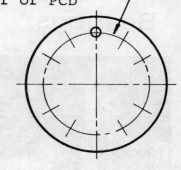

Holes on a *circular* pitch

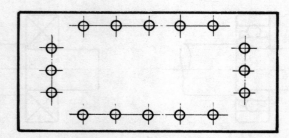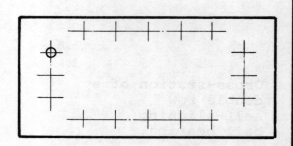

Holes on a *linear* pitch

42

Conventional representation: Gears (1)

Before gears can be drawn a great deal of background knowledge about their nomenclature and construction must be acquired. The following drawings of gears are presented as conventions only.

Subject - GEARS   *Convention*

Note

Side view of gear wheel is in section.

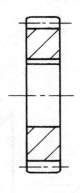

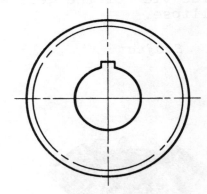

Single spur gear

SPUR GEAR

Note

A pinion is so called when it has a relatively small number of teeth compared with its mating gear wheel.

Side view

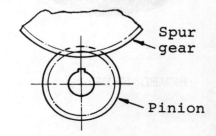

Spur gear

Pinion

In mesh with a pinion

WORM AND WHEEL

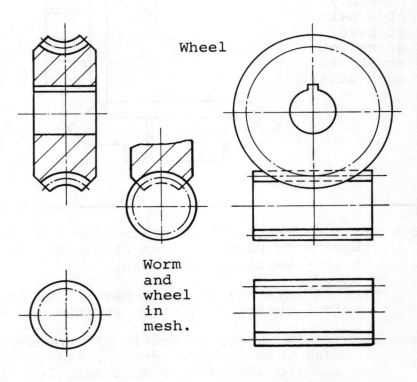

Wheel

Worm and wheel in mesh.

Worm

43

Conventional representation: Gears (2)

A good example of how a complex component may be drawn relatively simply is the bevel gear. The assembly shown below is of a pair of gears of equal size, the direction of motion being changed through an angle of 90°. In this arrangement, the gears are often referred to as mitre wheels.

The gears may be of differing sizes, of course, and the angle between the shafts may be other than 90°. In this latter case, the side view of the gear assembly would have to show one gear as an ellipse.

Subject   *Convention*

BEVEL WHEEL

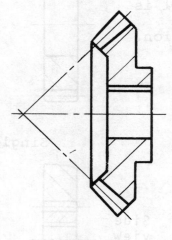

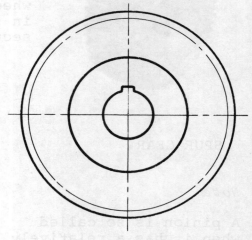

Assembly
of a pair
of bevel
gears set
at 90° to
each other.

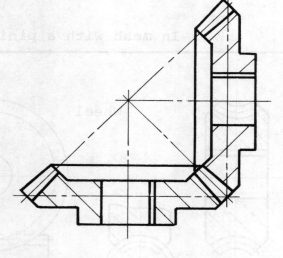

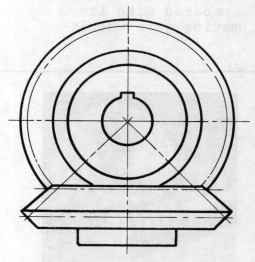

Note   BS 308: Part I: 1972 deals with a number of features which require to be shown less frequently than those illustrated in this section. They are:

Thread inserts; repeated parts; semi-elliptic springs; conical springs; torsion springs and a rack and pinion.

Summary   In all cases, conventional representation should only be used when it *saves drawing time*. Where it is not considered adequate, a more detailed view should be shown.

Test for terminology

Select, from the array, the conventional engineering term for each item shown. Insert in the answer column the corresponding number and letter.

Solutions are given on page 148.

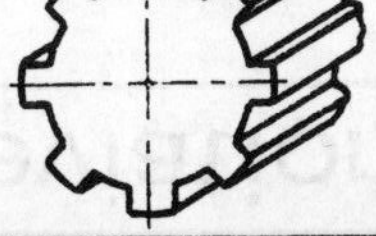

ARRAY

	A	B	C	D
1	Countersunk Hole	Undercut or Groove	"Blind" Hole	Serrated Shaft
2	Slot	Flange	Collar	Tongue
3	Shoulder	Boss	Keyway	Web
4	Taper	"Blind" Tapped Hole	Spigot	Flanged Bush
5	Bore	Spotface	Splined Shaft	Counterbore
6	Threaded Hole	Rib	Chamfered	Tube

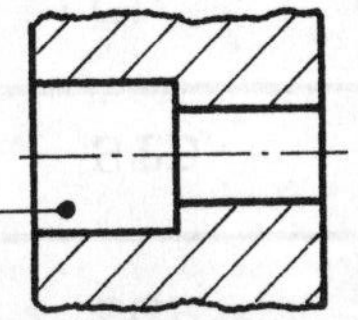

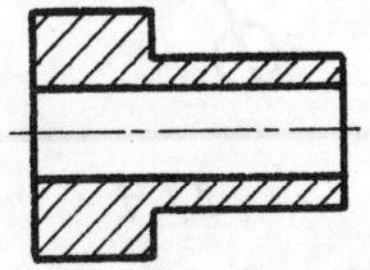

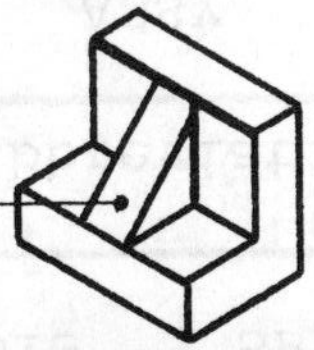

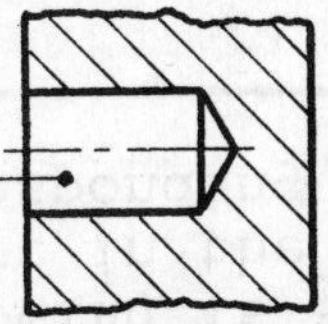

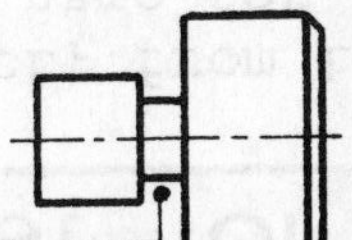

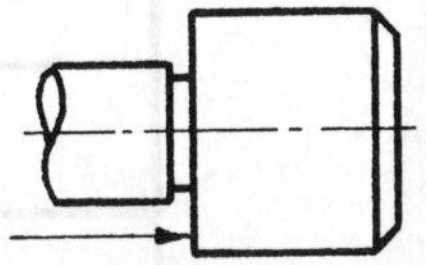

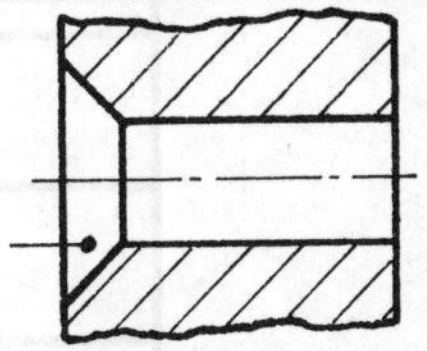

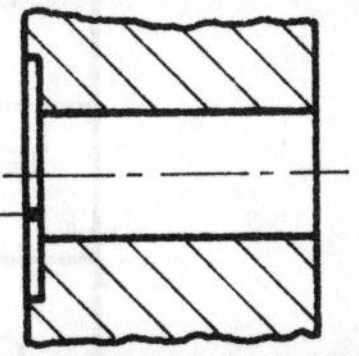

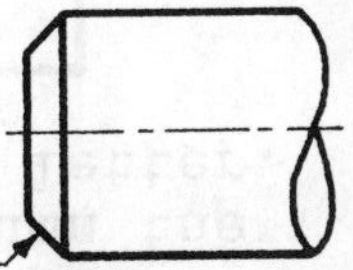

Example — 3C

?

6 ?

7 ?

8 ?

9 ?

10 ?

1 ?

2 ?

3 ?

4 ?

5 ?

Solutions on page 148

ARRAY

	A	B	C	D
1	Across Flats	Radius	Screw	Diameter (preceding a dimension)
2	Clearance	Spherical	Long	Drawing
3	Assembled	Stated	Across Face	Right Hand
4	Diameter (in a note)	Standard	Assembly	Chamfer
5	Placed	Specification	Pitch Circle Diameter	Centre Line
6	Chamfered	Drawn	Screwed	Light Gauge

Test for abbreviations

Select, from the array, the correct full name for which each of the following is an abbreviation. Insert in the answer column the corresponding number and letter.

Example SPEC 5 B

No.	Abbreviation	Answer
1	ASSY	
2	A/F	
3	Ø	
4	SCR	
5	STD	
6	CL	
7	R	
8	LG	
9	DRG	
10	DIA	
11	PCD	
12	CHAM	

Test for conventional representations

Select, from the array, the names of the items for which the following diagrams are conventional representations.
Insert in the answer column the corresponding number and letter.

Solutions on page 148

	Conventional reps	Answer
Example		4C
1		
2		
3		
4		
5		
6		
7		
8		
9		
10		

ARRAY			
	A	B	C
1	Worm Wheel	Shoulder Knurling	Bevel Gear
2	Tension spring	Round Shaft	Spur Gear
3	External Screw Thread	Splined Shaft	Break in Rectangular Shaft
4	A Resistor	Worm	Break in Tubular Shaft
5	Square on Shaft	Form of Machining Symbol	Compression spring
6	Machine All Over	Internal Screw Thread	Counterbore
7	Serrated Shaft	Diamond Knurling	Bearing
8	Weld symbol	Collar	Straight Knurling

Test for abbreviations, terminology and conventional representations

The sectional assembly drawing below has twenty different numbered items. Insert the correct conventional name for each numbered item in the table below. For example, item 2 is a centre line: this is shown below in space 2 in the table. Solutions on page 148

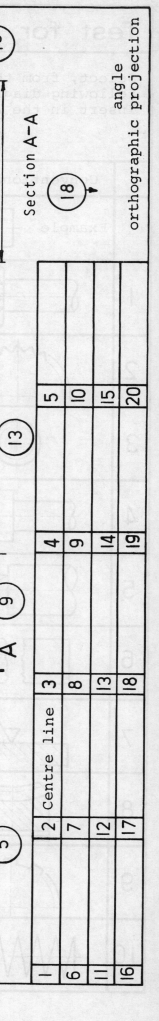

ø 15 mm

50mm CRS

Section A–A

Front Elevation

angle orthographic projection

1				
6				
11				
16				

2	Centre line	3		4		5	
7		8		9		10	
12		13		14		15	
17		18		19		20	

Pictorial Drawing

A component may be represented graphically in various ways.
An Orthographic Drawing, for example, requiring a minimum of
two views to fully communicate the size and shape of a component, is used in engineering mainly to convey manufacturing
instructions from the designer to the craftsman. On the
other hand a well executed Pictorial Drawing, adequately
representing all but the most complicated components using
one view only, is used mainly as an aid to visualization of
the shape of a component rather than for communicating
detailed instructions for manufacture.

A pictorial drawing, generally, is a quickly produced,
approximately scaled representation of a component - a
"picture" rather than an accurately scaled line drawing.

There are many different types of pictorial representation. Two of the most commonly used ones are known as
Isometric Drawing and Oblique Drawing both of which are discussed in detail in this section.

Note! There is a method of isometric representation, more
precise than the one discussed here, known as Isometric
Projection. This method is much more time-consuming, and
therefore less commonly used, because all sizes for the
drawings have to be transferred from a specially constructed
isometric scale. It is the approximate method of representation, Isometric Drawing, which is used in this section.

Pictorial Drawing: Isometric

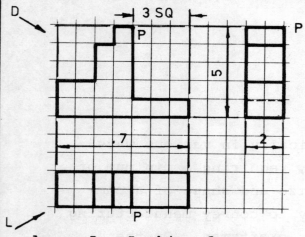

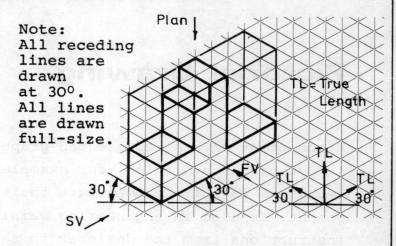

Note:
All receding
lines are
drawn
at 30°.
All lines
are drawn
full-size.

Arrow L - Looking from LEFT
Arrow D - Looking DOWN.

Isometric drawing

When making an isometric drawing from orthographic views of a component use an isometric grid at first, as shown above. Follow this simple procedure:

1. Choose the direction from which the component is to be viewed so that the resulting isometric drawing will show the most detail of the component. Compare the isometric drawings below.

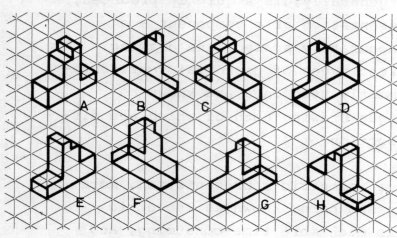

Example A is drawn looking from the left and down, and is a good representation of the component.

Example C shows the same detail but is drawn from a different direction.

All of the other drawings fail to show some of the details of the component.

2. Draw the outline of a box into which the component will just fit.

3. Use distances from the orthographic views to set out points within this box, e.g. P.

4. Heavy-in the lines when the outline is completed.

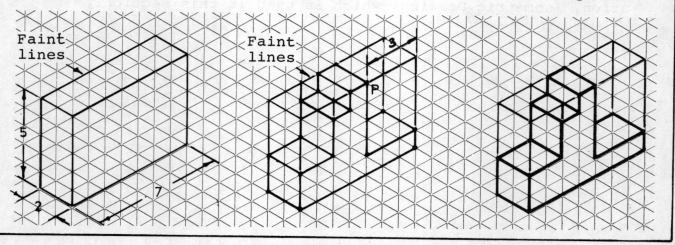

Faint lines

Faint lines

Isometric drawing: inclined edges

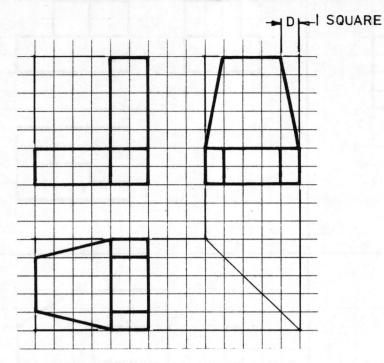

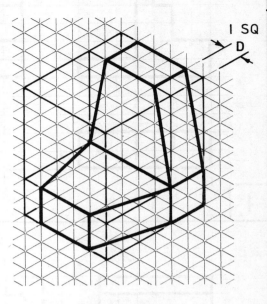

Isometric

When components have sloping surfaces, the edge distances in the orthographic views are transferred to the isometric box as shown by the dimension D in the above example.

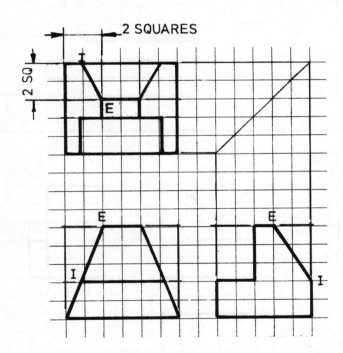

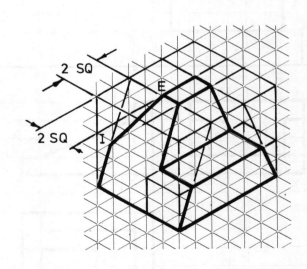

The component shown above has surfaces sloping in two directions resulting in inclined edges, for example, IE. In this case, pairs of distances from the orthographic views are used to plot points within the isometric box. Distances used in this way for location purposes are called ORDINATES.

Note:- The angles of inclination in the orthographic views are *never* transferred to the isometric box. When I and E have been located, they are joined to produce the correct slope in the isometric drawing.

Isometric drawing exercises

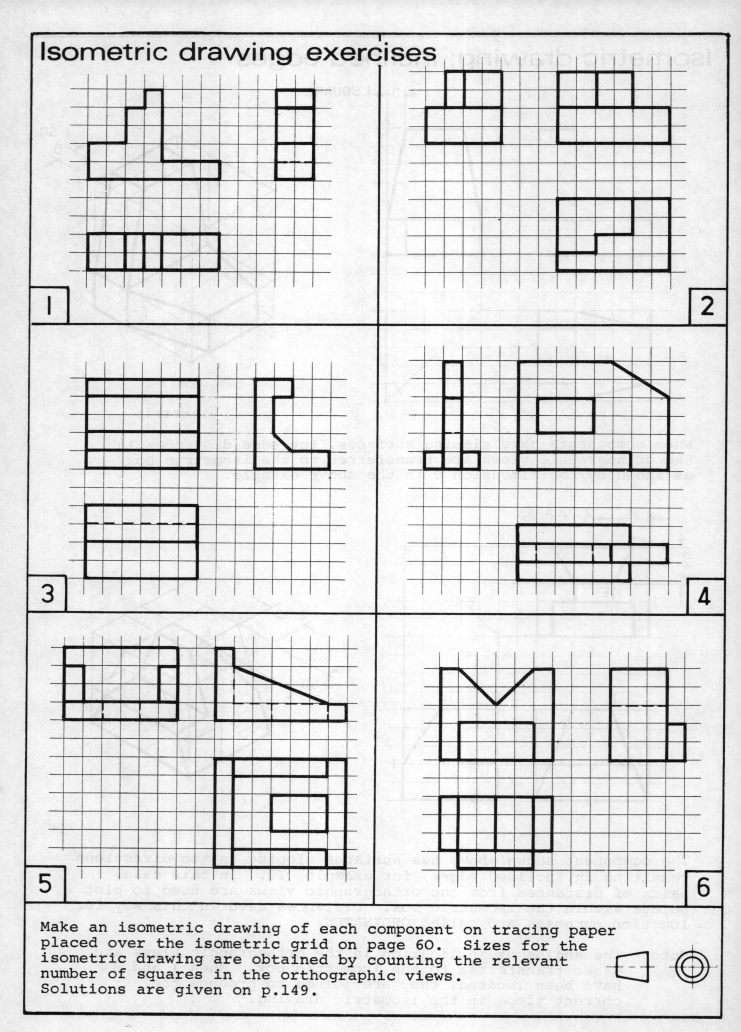

Make an isometric drawing of each component on tracing paper placed over the isometric grid on page 60. Sizes for the isometric drawing are obtained by counting the relevant number of squares in the orthographic views.
Solutions are given on p.149.

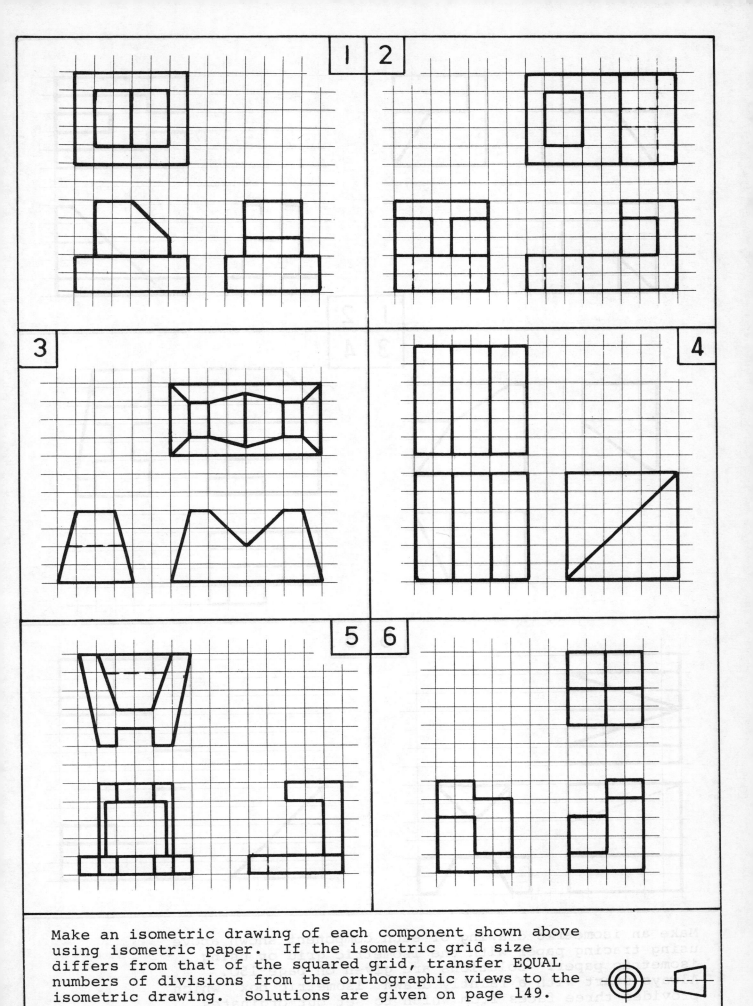

Make an isometric drawing of each component shown above using isometric paper. If the isometric grid size differs from that of the squared grid, transfer EQUAL numbers of divisions from the orthographic views to the isometric drawing. Solutions are given on page 149.

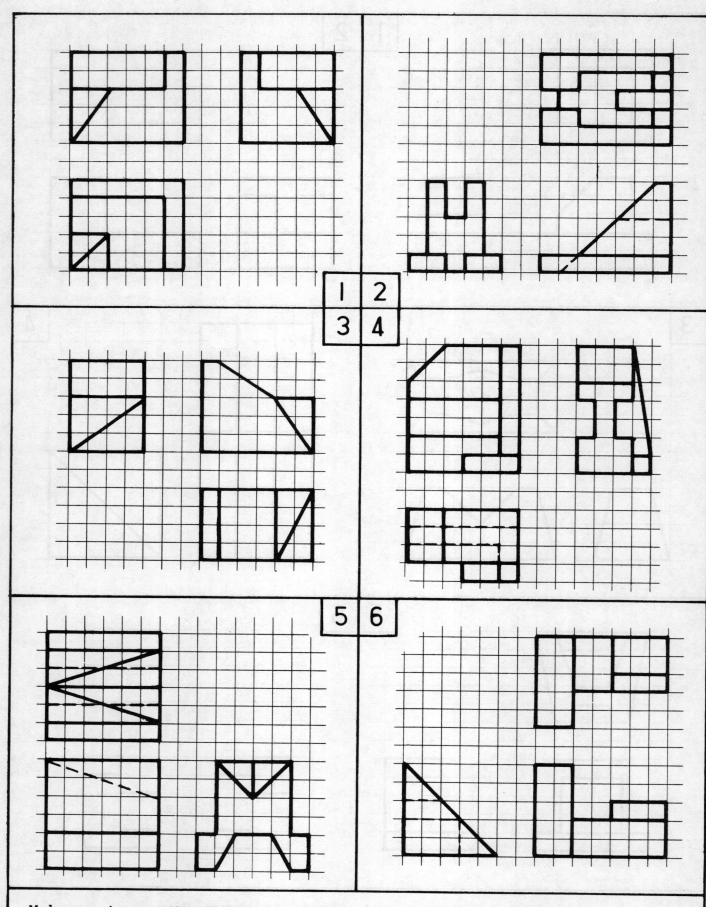

Make an isometric drawing of each component shown above
using tracing paper with the isometric grid or using
isometric paper. Solutions are given on page 150.
Always start with a faintly drawn "isometric box" which
provides three faces from which to set out ordinates.

54

Isometric drawing: construction of ellipses (1)

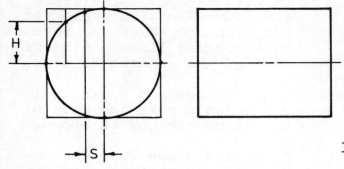

A shape which is circular on an orthographic view is shown as an ellipse on an isometric drawing.
The following constructions illustrate 3 different methods of producing the elliptical shape.

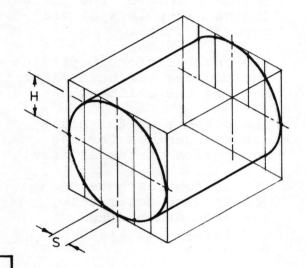

1. Lines, called ordinates, are drawn on the circle in the orthographic view as shown. Corresponding ordinates are drawn on the isometric drawing with the same spacing S. The height H of each ordinate on the true circle is transferred to the corresponding ordinate on the isometric drawing to form the outline of the ellipse. These points are joined with a neat freehand curve.
The more ordinates, the more accurate the ellipse.
As the circle and ellipse are symmetrical, the ordinates need only be constructed in a quarter-circle.

I

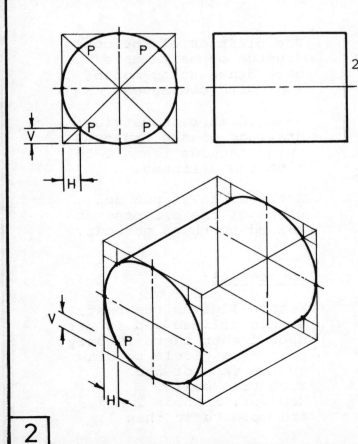

2. Diagonals are drawn on the orthographic side view as shown on the left.

Horizontal and vertical ordinates are drawn to touch the points where the circle crosses the diagonals.

The ordinates are transferred to the isometric box as shown and the resulting points of intersection P are joined with a neat freehand curve.

As in method 1, only a quarter-circle need be drawn to obtain the four points required.

This method is useful for quick freehand sketching but is not as accurate as method 1.

2

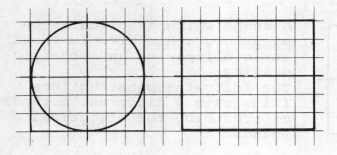

3. In this method, the ellipse is constructed on the isometric box with compasses and several construction lines are needed to find the centres of arcs.

Centre lines and a diagonal are drawn across the face of the isometric box as shown.

Lines are drawn from the bottom corner of the box B to the mid-points of the opposite sides.

The intersections C provide the centres for both minor arcs of the ellipse.

The corners of the box B are the centres for the major arcs of the ellipse.

If the centres have been constructed accurately, the arcs drawn with compasses will blend perfectly.

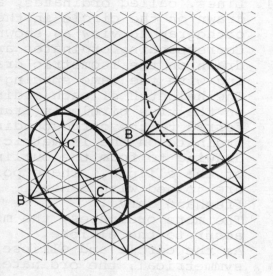

3

FRUSTUM
(a "beheaded" cone or pyramid)

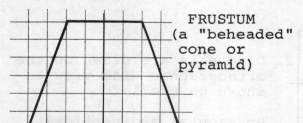

4. The ellipses of the conic frustum opposite have been drawn using method 3, i.e. with compasses.

Make further isometric drawings of the frustum using methods 1 and 2 to plot the ellipses.

Compare the shapes and sizes of the ellipses obtained by the 3 methods.

In general:

Method 1 gives the best shaped ellipse and is very useful when there is only part of a circle to draw on the face of the isometric box.
However, methods 2 and 3 are more rapid than 1.

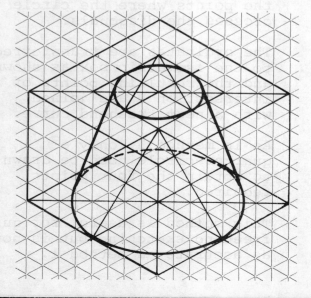

4

Isometric drawing examples

EXAMPLE 1

An isometric drawing
of the Vee-block shown
on page 21

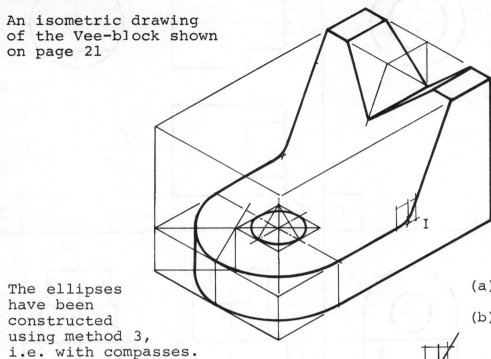

The Vee and the
sloping face
have been con-
structed by
using pairs of
ordinates as
shown on
page 51

The small
radius at the
intersection
of surfaces I
may be drawn:

The ellipses
have been
constructed
using method 3,
i.e. with compasses.

(a) using ordinates
or
(b) with a radius
gauge if the
curve is small.

EXAMPLE 2

The incomplete
isometric drawing of
the component below
is viewed from the
right and
looking down.

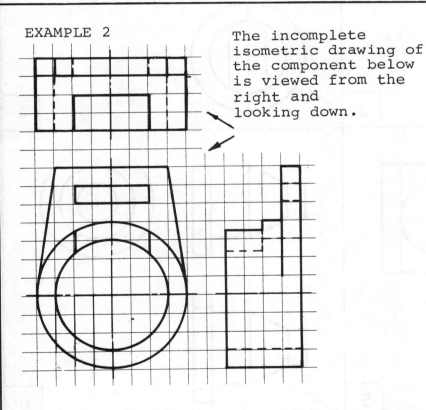

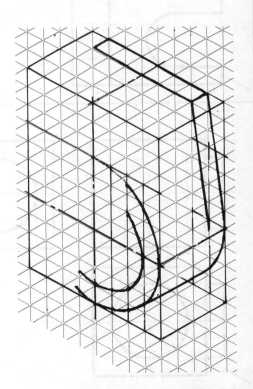

Use the procedure suggested on page 52 to complete the
isometric drawing in the space provided or make a complete
isometric drawing on isometric grid paper. For further
examples use the drawings provided for sectioning exercises
on pages 28-31 (solutions are not given).

Isometric drawing exercises

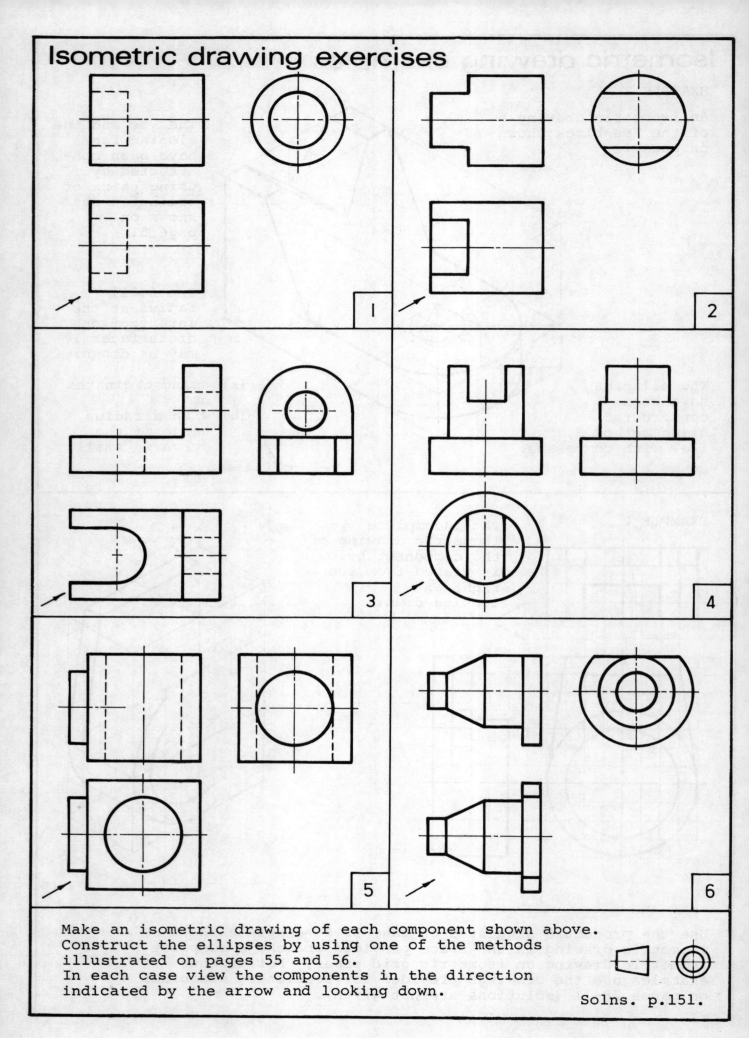

Make an isometric drawing of each component shown above.
Construct the ellipses by using one of the methods
illustrated on pages 55 and 56.
In each case view the components in the direction
indicated by the arrow and looking down.

Solns. p.151.

Isometric freehand drawing

Each drawing on this page consists of two views of a component in First Angle ortho-graphic projection.

From these views, sketch an isometric view of each component in the space provided.

Hidden detail need not be shown.

An example is given below

Solns. p.152.

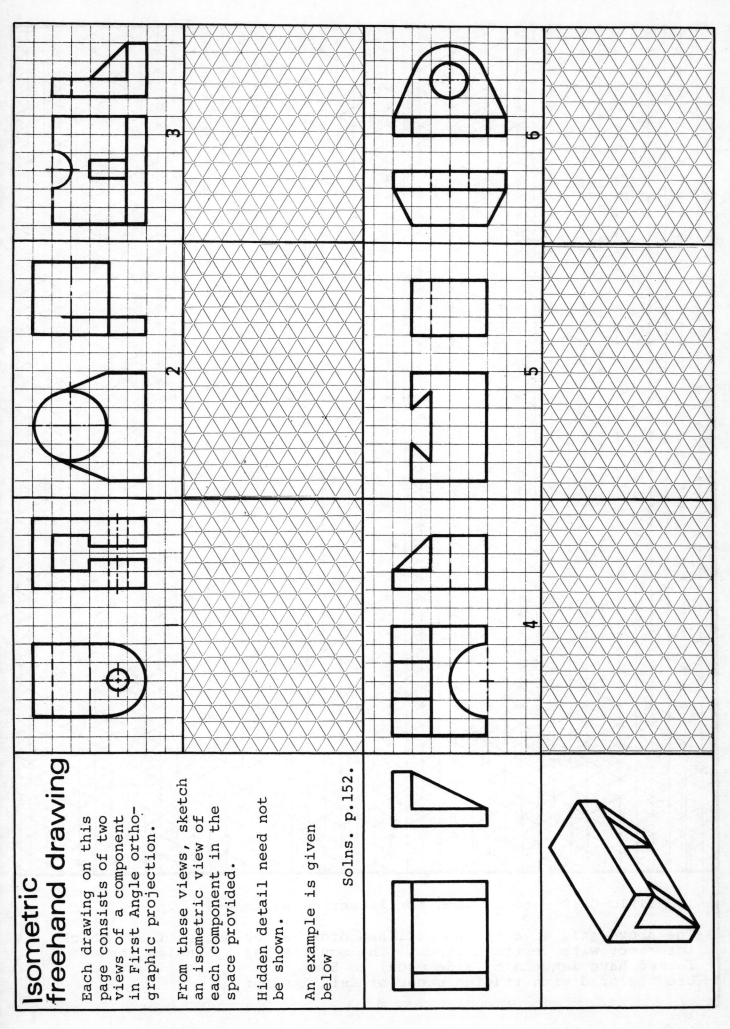

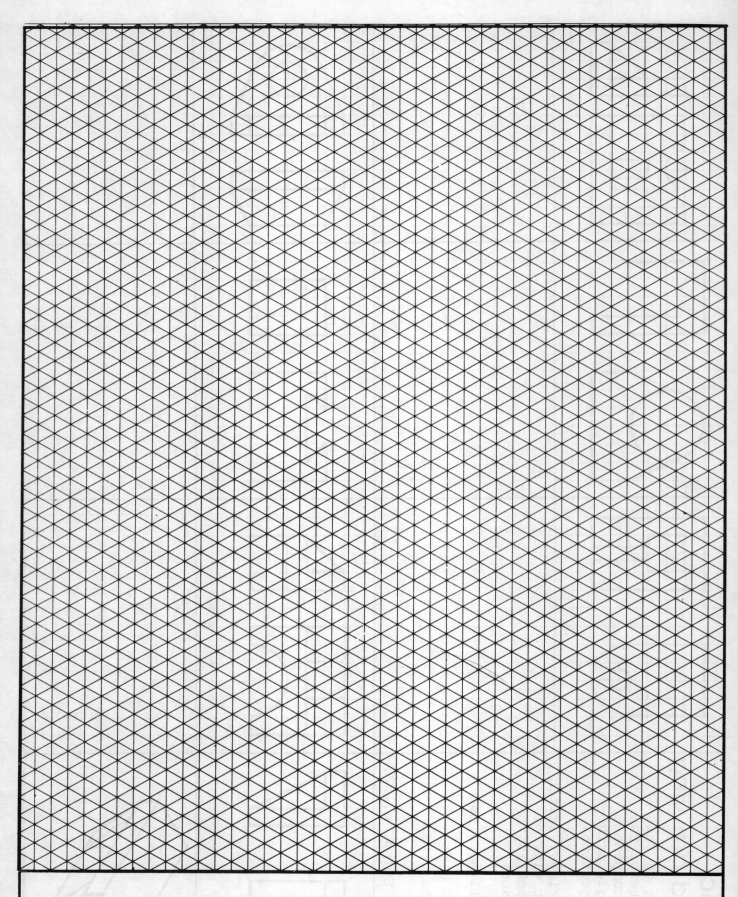

ISOMETRIC GRID - to be used for isometric sketches.

The above grid is composed of lines drawn at an angle of 30° which intersect with vertical lines. The equilateral triangles thus formed have lengths of side equal to 5 mm.
Use the grid with tracing paper or detail paper.

Pictorial Drawing: Oblique

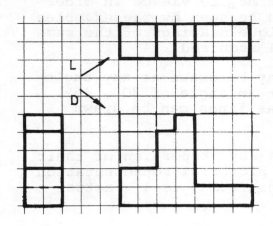

Arrow L - Looking from LEFT
Arrow D - Looking DOWN.

An oblique pictorial drawing presents the component with one of its faces as a true shape. This true shape is drawn on the front face of the oblique box as shown below.

The longest face is usually drawn on the front of the oblique box with receding lines between ½ and ¾ full size.

Compare these oblique drawings. Drawings 7, 8 and 9 represent the shape of the component better than the others.

Note:- **TL** = TRUE LENGTH

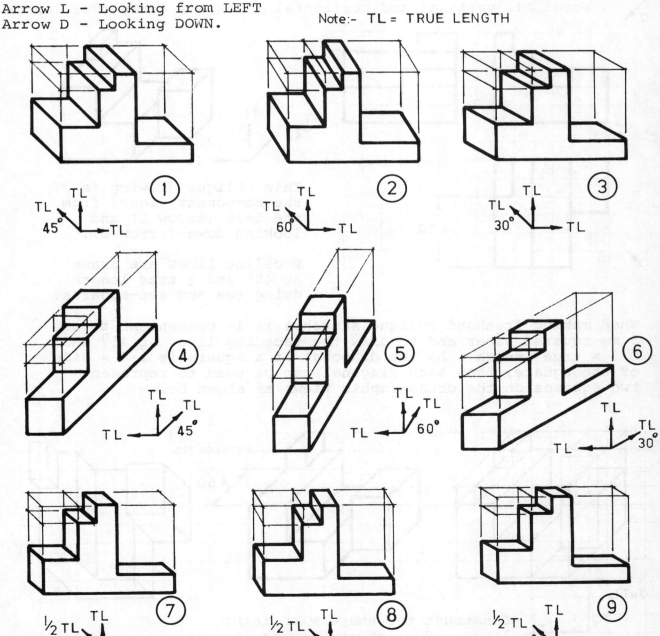

Oblique drawing: methods of construction

There are many variations in angle, length of receding lines, and directions from which a component may be viewed in order to produce an oblique drawing, as can be seen by the examples on the previous page. Different oblique drawings of the same component may each provide the detail required.

In general:

1. The receding lines may be drawn at any angle to the horizontal but an angle of 30°, 45° or 60° is preferred as lines can be drawn with set-squares.

2. Receding lines may be any proportion of their true length. A good pictorial representation is obtained if lengths from ½ to ¾ × actual length is used.

Note: All vertical and horizontal lines are drawn true length.

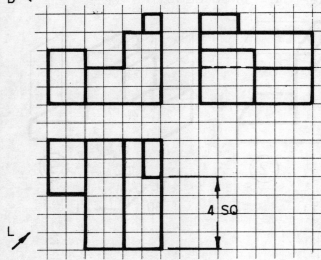

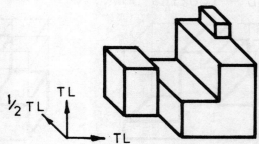

This oblique drawing is of the component viewed from the left (arrow L) and looking down (arrow D).

Receding lines are drawn at 45° and ½ true length using tee and set-squares.

When making freehand oblique sketches it is convenient to use 5 mm squared paper and to draw the receding lines at 45° and 0·7 × true length. As the diagonal of a square is 1·4 × side of the square, then each diagonal can be used to represent two squares on the orthographic views as shown below.

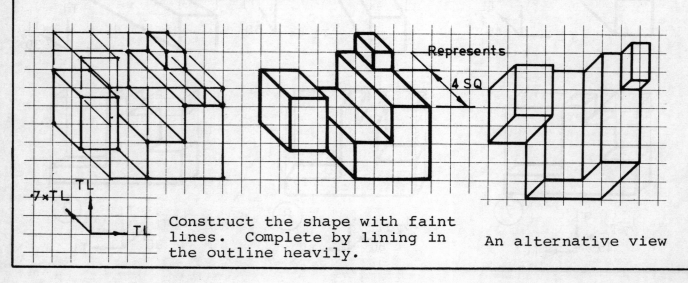

Construct the shape with faint lines. Complete by lining in the outline heavily.

An alternative view

Oblique drawing: inclined edges

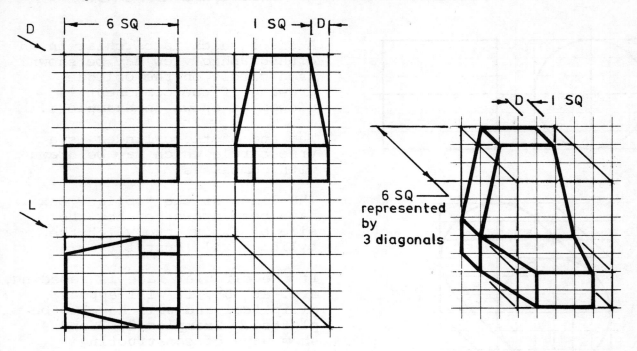

When components have sloping surfaces, the edge distances in
the orthographic views are transferred to the oblique box as
shown by the dimension D in the above example.

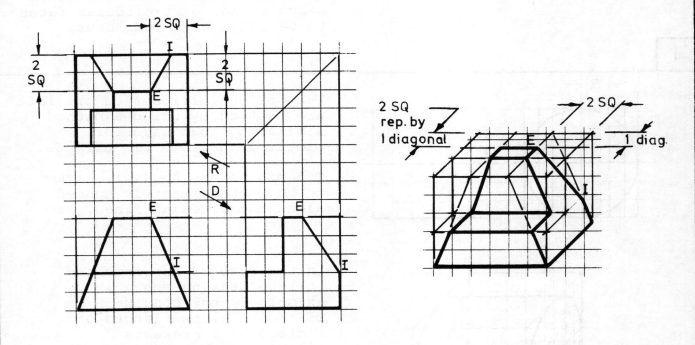

The component shown above has surfaces sloping in two directions
resulting in inclined edges, for example IE. Pairs of ordinates
from the orthographic views are used to plot points within the
oblique box, for example point E.

Note: The angles of inclination in the orthographic views are
        never transferred to the oblique box. When I and E
        have been located they are joined to produce the correct
        slope in the oblique drawing.

Methods of constructing "oblique circles"

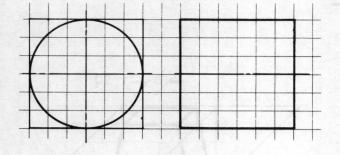

A shape which is circular in an orthographic view may be shown as a circle on the oblique drawing if the component is viewed as shown in example 1.

The end of the cylinder is a true circle which may be drawn with compasses.

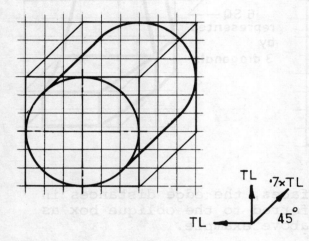

The lines of the "oblique box" have been drawn to show the size of a box into which the cylinder will just fit.

If the circular face is receding, e.g. at 45°, it will appear elliptical and will have to be plotted. Examples 2, 3 and 4 show ways of constructing ellipses.

When making an oblique drawing view the component, if possible, in a direction which will allow circular and part-circular faces to be drawn with compasses.

1

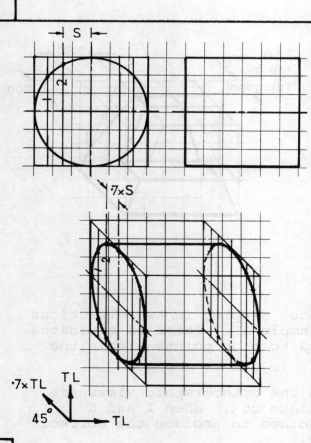

The oblique ellipses on the left are plotted by using a method similar to that used for plotting an isometric ellipse.

The ordinates, however, have to be proportionately spaced along the receding lines.

In the case of the freehand oblique drawing shown in example 2, the spacing S is O·7 × S i.e. the length of 1 diagonal represents 2 squares on the orthographic view.

As in the construction for an isometric ellipse, the ordinates need only be constructed in a quarter-circle.

2

Methods of constructing "oblique ellipses"

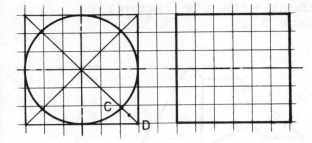

The method for constructing an oblique ellipse shown opposite is only used when the receding lines are drawn at 0·7 × true length, mainly for quickly drawn freehand sketches.

The larger scale drawing below shows how the length CD is obtained. The length ½CD can be judged by eye.

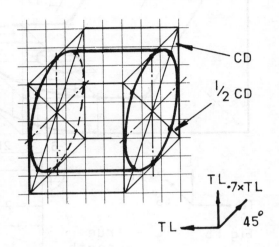

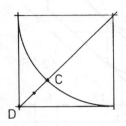

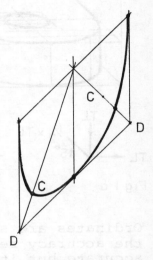

Note: The actual length of CD is transferred to the oblique box.

3

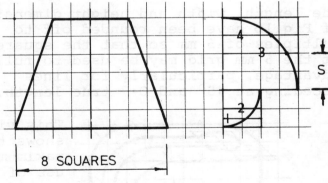

8 SQUARES

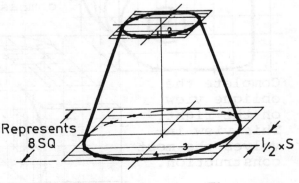

Represents 8 SQ

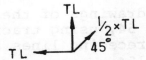

The ellipses of the conic frustum in example 4 have been drawn freehand using the ordinate method.

The ordinates used for plotting the points on the ellipse have been measured along horizontal lines spaced proportionately.

Ordinates are true lengths.

Spacings are ½ × true length.

The ordinate method is also useful when there is only part of a circle to be drawn in the oblique box.

Exercises: Make oblique drawings of the components shown on page 58. Solutions are not given.

4

Oblique drawing examples

EXAMPLE 1   Two oblique drawings of the Vee-block shown orthographically on page 21.   Compare these with the isometric
drawing of the same component on page 57.

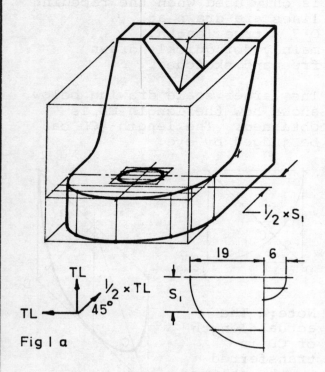

Fig 1 a

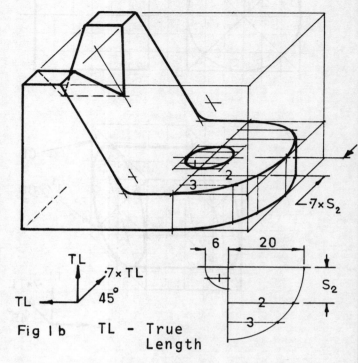

Fig 1 b   TL - True
          Length

TL - True Length

Ordinates are spaced to suit the size of ellipse being plotted and
the accuracy required.   More ordinates will make the drawing more
accurate but it will take longer to draw.   Note that ordinates set
off from receding lines are spaced at proportionate distances along
these lines.

Fig.1a has been drawn
using the actual dimensions
of the vee-block.   Receding
lines at 45° and ½ TL.

EXAMPLE 2

The length, width and height dimensions
of Fig.1b have been rounded off to
multiples of 5 mm so that the squares
of the 5 mm grid may be used to full
advantage, particularly for lines
receding at 45° and 0·7 true length.

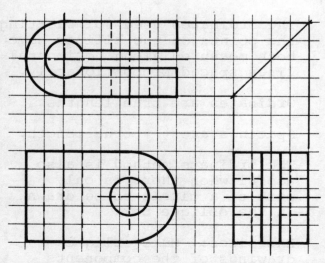

This view
shows the
maximum
use of
compasses.

Complete the
oblique view
on the right.→
This view in-
volves the most
construction.

For further exercises: Make oblique drawings of the
components shown on pages 52, 53 and 54, using tracing paper and the
5 mm squared grid on page 67.   Draw receding lines at 45° and 0·7 × TL.
Solutions are given on pages 153 and 154.

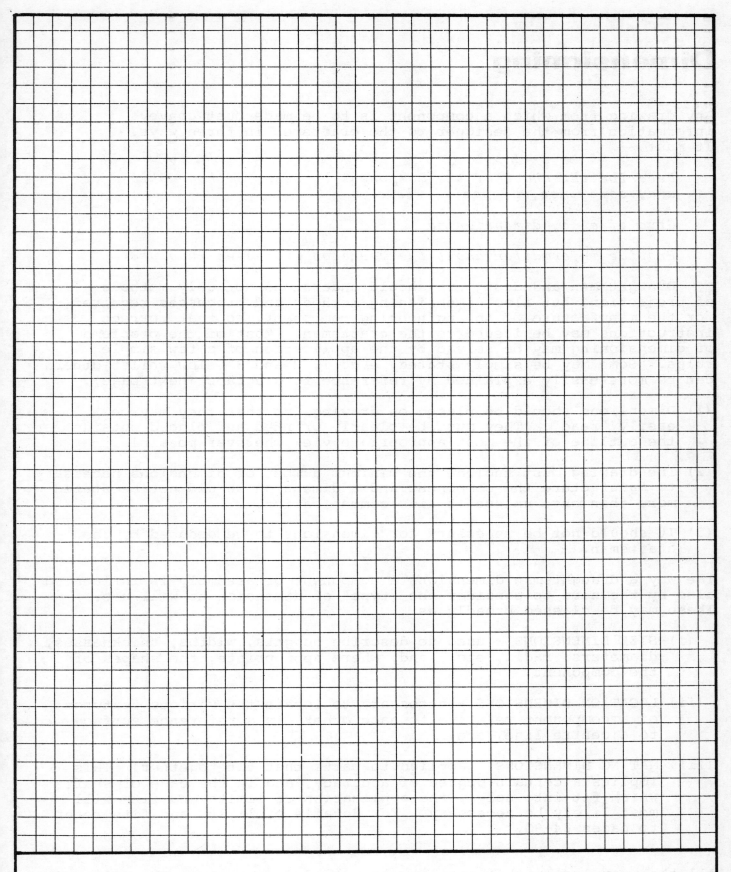

SQUARED GRID - to be used for oblique pictorial drawing and orthographic drawing and sketches.

The above grid is composed of lines drawn to form squares which have
lengths of side equal to 5 mm.
Use the grid with tracing paper or detail paper.

Dimensioning

Any drawing from which a component is to be made must convey
information from the designer to the craftsmen in three ways.
It must

> *Describe the shape of the component by using orthographic
> and, sometimes, pictorial views*
>
> *Give sizes by dimensioning*
>
> *Provide information about the workshop processes involved*

Draughtsmen and designers should understand not only methods
of projection, but also dimensioning methods and processes required
for the manufacture of an engineering component, so that correct
instructions may be issued to the craftsman. Most of the problems
of dimensioning may be solved by the application of a few simple
rules. Some may be simply stated, e.g. 1, 2 and 3 below whilst others
can be more easily explained by reference to diagrams (next page).

(1) Dimensions should be placed on drawings so that they may be
easily "read". They must be clearly printed and placed outside
the outline of the most appropriate view wherever possible.

(2) The drawing must include the minimum number of dimensions necessary
to manufacture the component and a dimension should not be stated
more than once unless it aids communication.

(3) It should not be necessary for dimensions to be deduced by the
craftsman.

The way a draughtsman dimensions a component or assembly is influenced
also by the need to consider the "type" of dimension he will use.
These may be classed broadly as

(i) SIZE DIMENSIONS - used to describe heights, widths, thicknesses,
diameters, radii, etc., and, where appropriate, the shapes of
the component.

(ii) LOCATION DIMENSIONS - necessary to locate the various features
of a component relative to each other, to a reference surface or
to a centre line, etc.

(iii) MATING DIMENSIONS - applied to parts that fit together. This
implies a certain degree of accuracy, and in the case of shafts
which fit into holes, the application of limits and fits will
most likely be necessary. For details of limits and fits refer
to pages 84-90.

All the examples in this section are intended to explain the rules of
dimensioning which are set out as recommendations in BS 308:Part 2:1972.
It must be stressed, however, that the dimensioning of any component
depends on the draughtsman's interpretation of the recommendations.
A component may be dimensioned in a number of different ways yet still
be dimensioned correctly.

Dimensioning

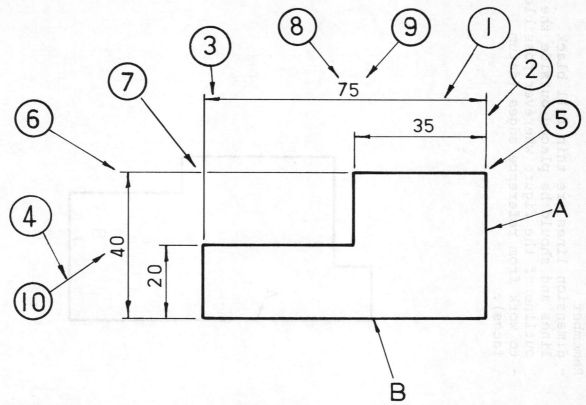

A number of the basic rules of dimensioning can be explained by reference to the above drawing of a thin plate.

The sides marked A and B are known as DATUM faces. They are used as reference edges from which dimensions are drawn. Datums may or may not be machined. Even if they are not machined it is good practice to choose reference edges in order to simplify the layout of dimensions.

Points to note:

(1) DIMENSION LINES - thin full lines placed outside the component wherever possible and spaced well away from the outlines. The longer dimension lines are placed outside shorter ones.

(2) PROJECTION LINES - thin full lines which extend from the view to provide a boundary for the dimension line. Drawn at 90° to the outline.

(3) ARROWHEADS - drawn with sharp strokes which must <u>touch</u> the extension lines.

(4) A LEADER LINE is a thin full line which is drawn from a note, a dimension or, in this case, a "balloon" and terminates in an arrowhead or a dot.

(5) Relatively small gap.

(6) Relatively short tail.

(7) Crossing extension lines - usually a break to ensure clarity.

(8) Dimension placed above the dimension line. This is preferred to the alternative method of placing the dimension in a gap in the line. Avoid using both methods on the same drawing if possible.

(9) Dimension placed so that it may be read from the bottom or

(10) Dimension placed so that it may be read from the right hand side of the drawing sheet.

Study this information before turning to the next page.

The figure below is badly dimensioned.
How many dimensioning errors can you find?
Write your answer in the square on the right.

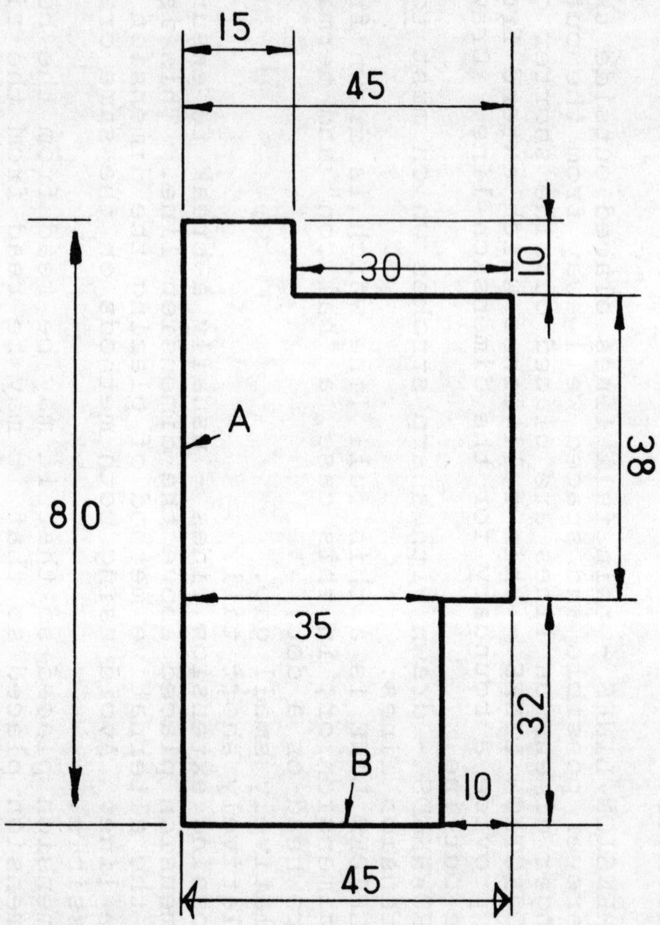

A and B are reference edges (datum faces).

Dimension the figure below correctly.
Remember
- dimension lines are thin full black lines and should be placed outside the outline of the figure wherever possible.
- to work from reference edges (datum faces).

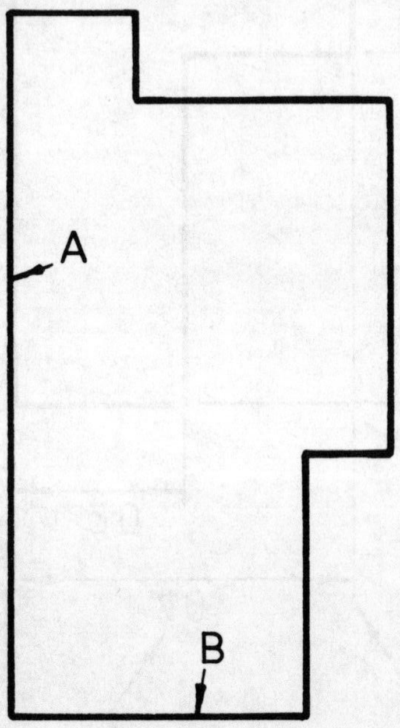

What is the least number of dimensions needed to dimension the plate with reference to edges A and B? (See p.71.)

Correct solution drawn FULL SIZE

Fig.1

Least number of dimensions 6

45

15

80

70

32

35

B

A

Plate 3 mm thick

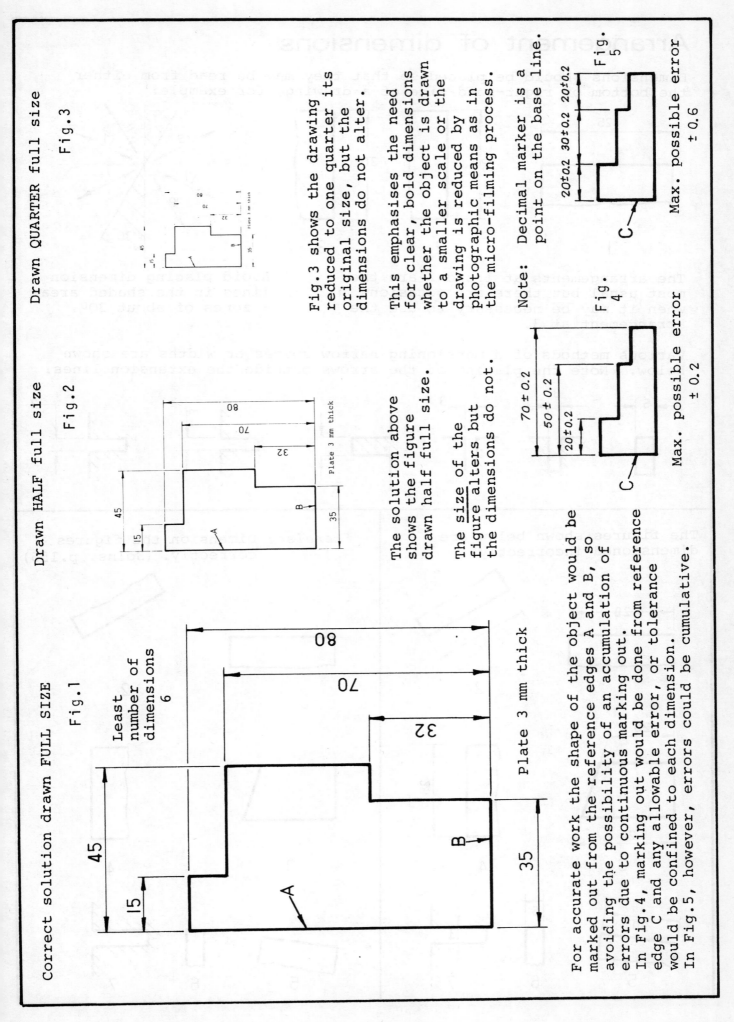

Drawn HALF full size

Fig.2

45

15

80

70

32

35

B

A

Plate 3 mm thick

Drawn QUARTER full size

Fig.3

45

15

80

70

32

35

B

A

Plate 3 mm thick

Fig.3 shows the drawing reduced to one quarter its original size, but the dimensions do not alter.

This emphasises the need for clear, bold dimensions whether the object is drawn to a smaller scale or the drawing is reduced by photographic means as in the micro-filming process.

Note: Decimal marker is a point on the base line.

The solution above shows the figure drawn half full size.

The size of the figure alters but the dimensions do not.

70 ± 0.2

50 ± 0.2

20 ± 0.2

C

Max. possible error ± 0.2

Fig. 4

20 ± 0.2   30 ± 0.2   20 ± 0.2

C

Max. possible error ± 0.6

Fig. 5

For accurate work the shape of the object would be marked out from the reference edges A and B, so avoiding the possibility of an accumulation of errors due to continuous marking out.
In Fig.4, marking out would be done from reference edge C and any allowable error, or tolerance would be confined to each dimension.
In Fig.5, however, errors could be cumulative.

Arrangement of dimensions

Dimensions should be placed so that they may be read from either the bottom or right-hand side of a drawing, for example:

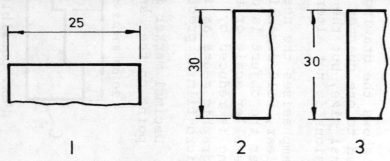

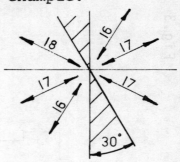

The arrangements at 1 and 2 are the most usual but there are occasions when it may be necessary to use the arrangement at 3.

Avoid placing dimension lines in the shaded area - zones of about 30°.

Various methods of dimensioning narrow spaces or widths are shown below. Note the placing of the arrows outside the extension lines.

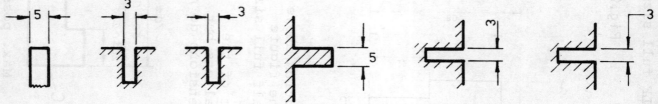

The figures shown below are dimensioned incorrectly.

Exercise: Dimension the figures correctly. (Solns. p.155)

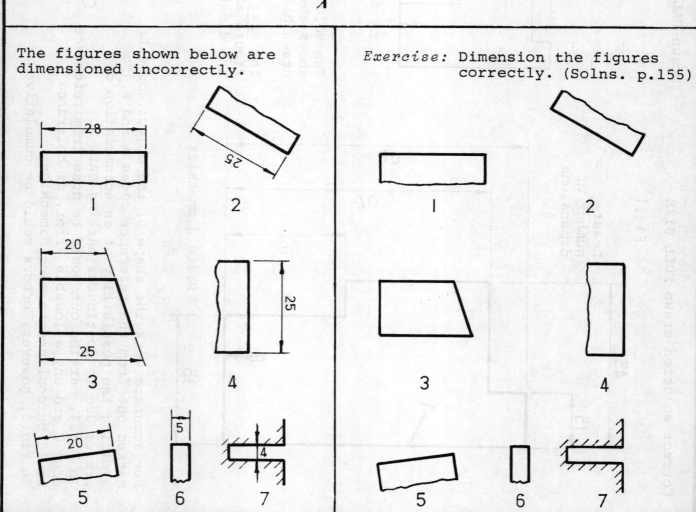

Dimensioning circles

On an engineering drawing a circle may be one of the following:

 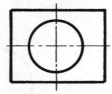 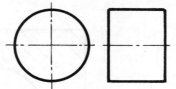 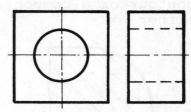

| A thin | A hole in a | The side view | The side view of a |
| disc | thin plate | of a cylinder | cylindrical hole |

The way a circle is dimensioned is influenced by the factors shown above and also by the size of the circle and the space available within the circle.

Points to note: (i) The dimension always refers to the diameter and NOT the radius
(ii) A circle is <u>never</u> dimensioned on a centre line.
(iii) The conventional symbol for diameter is ϕ

Methods used for dimensioning relatively small circles

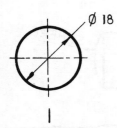  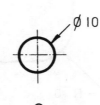

1   2   3   4

In 3 and 4 the leader line must be drawn in in line with the centre of the circle

and larger circles

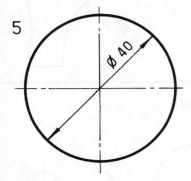 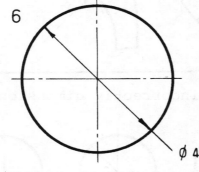

5   6

When the dimension line has to be drawn as in example 6 it is preferable to place the dimension as shown so that it may be easily read from the bottom of the sheet.

For diameters of cylinders

7

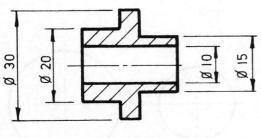 

Side view   Front view

In this example it is preferable to dimension the side view even though the cylindrical shape is not apparent. Dimensions in this view, however, must always be preceded by the symbol ϕ

Dimensioning radii

On an engineering drawing a radius usually describes the shape or contour of a component in a particular view and may be either

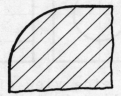 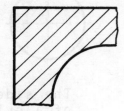 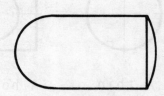

An external radius   An internal radius   or   A spherical radius

A radius should be dimensioned by a dimension line which passes through, or is in line with, the centre of the arc.
The dimension line should have one arrowhead which should be placed at the point of contact with the arc.
The abbreviation R should always precede the dimension.

The above statements may be interpreted as follows:

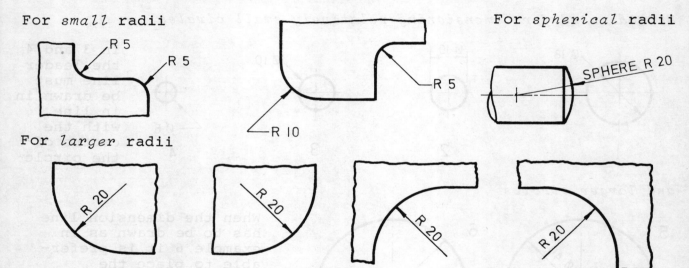

For *small* radii

For *larger* radii

For *spherical* radii

The circles shown below are incorrectly dimensioned.

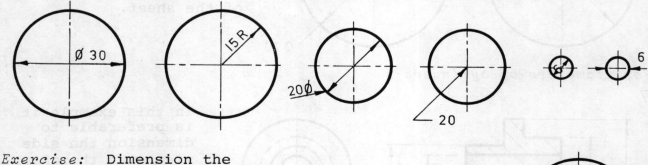

Exercise: Dimension the circles on the right correctly using the information given on the previous page (Solns. p.155)

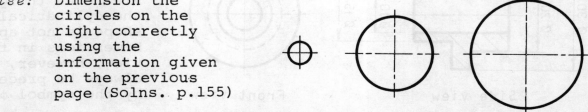

Dimensioning angles

Angles should be expressed in: (1) degrees   e.g. 90°
          or   (2) degrees and minutes e.g. 27° 30'
                                   0° 15'

The placing of the angular dimension depends on:

the *position* of the angle in relation to the bottom and/or the right-hand side of the drawing sheet and the *size* of the angle.

Angles may be dimensioned using one of the methods shown in these examples

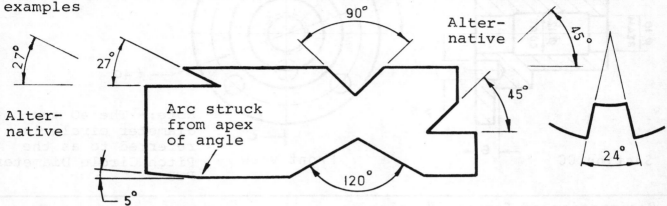

DIMENSIONING CHAMFERS

45° chamfers should be specified by one of the methods shown below:

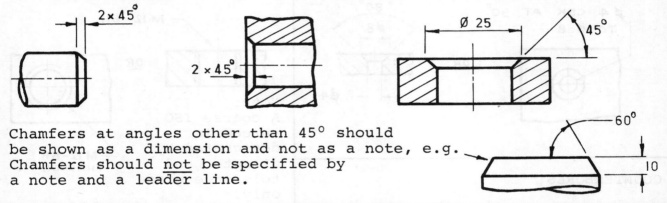

Chamfers at angles other than 45° should be shown as a dimension and not as a note, e.g. Chamfers should <u>not</u> be specified by a note and a leader line.

The radii below are dimensioned incorrectly.

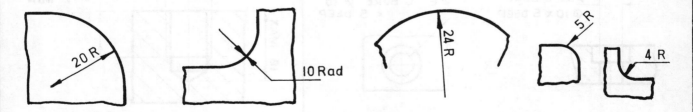

Exercise:- Dimension the radii below correctly using the information given on the previous page. (Solutions on page 155)

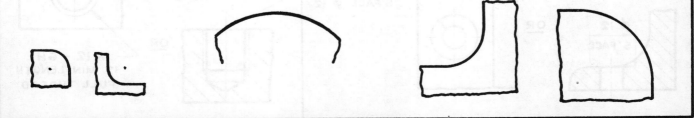

Designation of plain holes

It is often necessary to specify not only the diameter of a hole but also, the way the hole is produced. Processes for forming holes include drilling, boring, reaming and coring (holes made when casting). The drawing below shows how these may be designated.

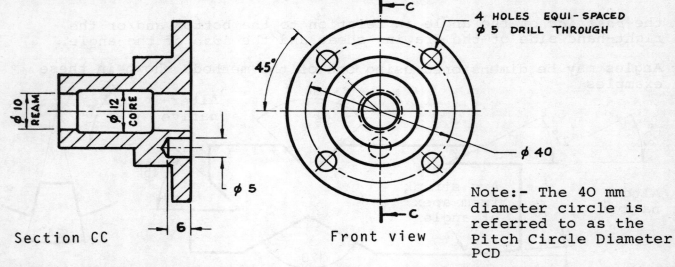

Section CC

Front view

4 HOLES EQUI-SPACED ⌀ 5 DRILL THROUGH

⌀ 40

Note:- The 40 mm diameter circle is referred to as the Pitch Circle Diameter PCD

Designation of "special" holes

COUNTERSINKS

⌀ 4 CSK AT 90° TO ⌀ 8

OR

90°
⌀ 8
⌀ 4

SCREW THREADS

M 12 - 6H

OR

Note:
A coarse ISO thread is designated by diameter and tolerance grade only.

M 12-6H
THROUGH

COUNTERBORES

⌀ 6 C'BORE ⌀ 10 x 5 DEEP

⌀ 6 C'BORE ⌀ 10 x 5 DEEP

OR

M 12-6H
18 MAX
12 MIN

OR

SPOTFACES

⌀ 12 S'FACE

OR

⌀ 6 S'FACE ⌀ 12

M 12 - 6H
12 MIN LENGTH
FULL THREAD

OR

M 12 - 6H
12 MIN LENGTH
FULL THREAD

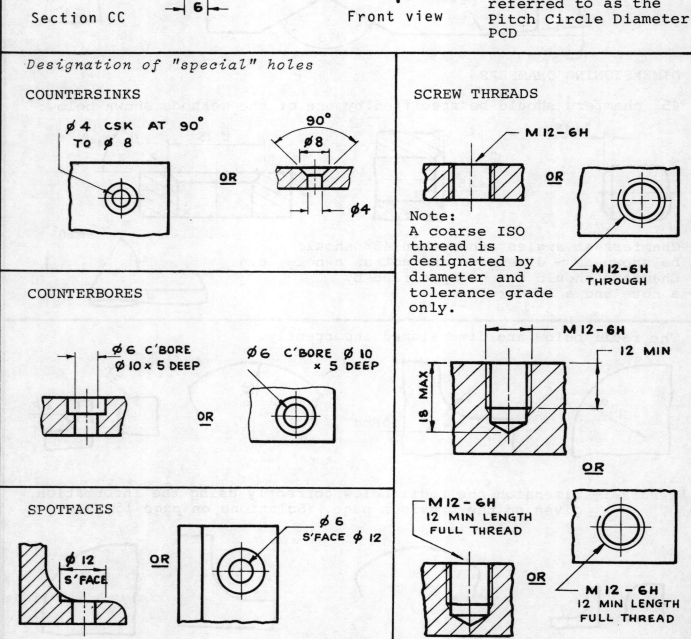

Location dimensions

Examples on previous pages have shown how components and features may be dimensioned when size is the main consideration.
Figures 1, 2 and 3 show how to dimension a component when location is necessary. The features can be located from a machined surface or centre line. Such a surface or line is known as a DATUM.

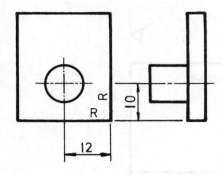

1.
Spigot located from two reference edges (R).

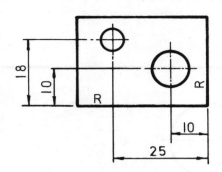

2.
Both holes located from two reference edges (R).

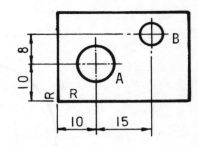

3.
Hole A located from two reference edges (R) then hole B related to hole A.

Size and location dimensions and use of the machining symbol

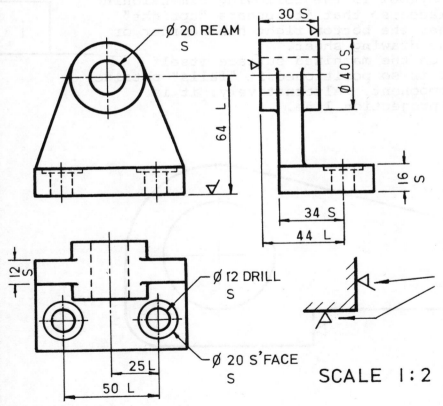

SCALE 1:2

4.
The simple bearing bracket casting on the left shows both size and location dimensions.

Reference surfaces are marked with a machining symbol:- ▽

This is placed so that it may be read from the bottom or from the right of the sheet.

It is preferable to place the symbol on the appropriate projection line rather than as shown on the left.

No symbol is required where the machining is specified i.e. in the case of the drilled holes, the reamed holes and the spot-face.

The location dimensions are those shown with a letter L and size dimensions by a letter S. Some of the size dimensions are less accurate than others e.g. the thickness of the rib is fixed during the casting process whilst the 20 mm diameter hole is accurately reamed. The 20 mm diameter reamed hole is <u>located</u> by the dimension from the machined base to the centre line of the hole.

Dimensioning exercises

Each drawing below shows a component made from mild steel plate
15 mm thick. Dimension both drawings so that accurate marking out
and subsequent manufacture of the plates may be carried out.
Use the machined surfaces as reference edges. The holes are to
be reamed. Sizes may be obtained by measuring the drawing.
All dimensions in millimetres.

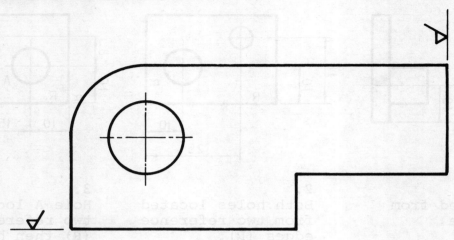

Note: Each machining symbol in the following dimensioning
exercises has been placed so that it appears "upright"
when viewed from either the bottom right hand corner or
right hand side of the drawing sheet.
The symbol is placed on the machined surface itself
provided that when it is so positioned it "falls" outside
the outline of the component. Alternatively, it is
placed on a suitable projection line.

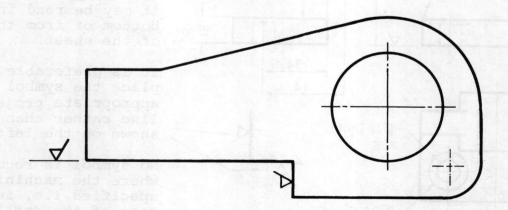

SCALE 1:2

Solutions to the above exercises are given on page 156.

78

Dimensioning exercise

Fully dimension this drawing of a cast iron Vee-block so that it could be manufactured if required. Add all necessary notes.

The hole in the base is to be drilled and the area around the top of the hole is to be spot-faced over a diameter of 20 mm.
The angle of the Vee is to be 90°.

Sizes may be obtained by measuring the drawing.

All dimensions in millimetres.

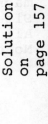

Solution on page 157

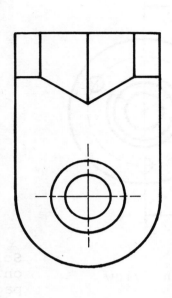

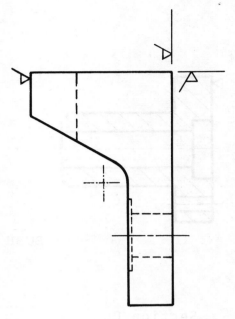

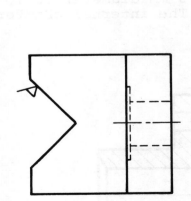

Dimensioning exercises

Fully dimension this drawing assuming that the component is to be machined all over. Take the datum as the face of the flange marked ✓ Sizes may be obtained by measuring the drawing.
All dimensions in millimetres.
Note: Holes   - 4 mm drill equi-spaced
       Chamfer - 2 mm at 45°

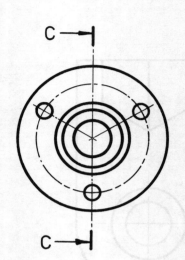

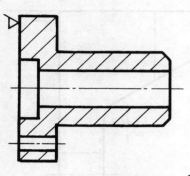

BUSH

Section CC   Front view   Solution on page 158   **4**

Fully dimension this drawing assuming that the component is to be machined all over. Take the datum as the face of the flange marked ✓
Note: (1) Use the horizontal centre line for locating the countersunk holes in the front view
      (2) The countersink is 90° and enlarges the drilled holes from 5 mm diameter to 10 mm diameter.
      (3) The internal chamfer - 3 mm × 60°

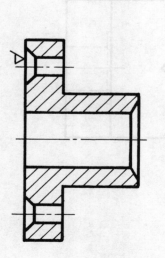

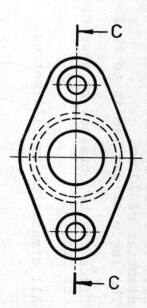

GLAND

Section CC   Front view   Solution on page 158   **5**

Dimensioning exercises

The component below is to be made of brass. Fully dimension the drawing to enable the bracket to be manufactured.

Note:- (1) The only parts to be machined are the back face (which should be used as the datum) and the tapped holes.
(2) Locate the holes from the centre line in the front view.
(3) Locate the holes and slot from point X in the side view.

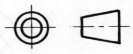

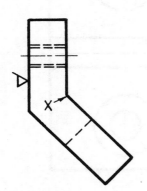

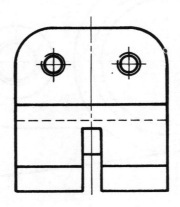

LOCATING
BRACKET

Side view

Front view

Solution
on
page 159

6

Fully dimension the drawing so that the adjusting screw may be manufactured. Take the shoulder marked ▽ as the datum face.

Note:- The thread is M 6. The head is knurled (fine diamond).
The chamfers - 2 mm at 45°
The radius at the end of the screw is 8 mm.

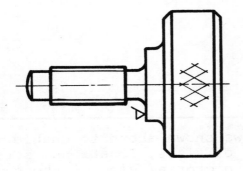

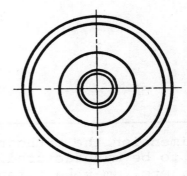

ADJUSTING
SCREW

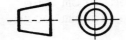

Front view

Machined
all
over

Side view

Solution
on
page 159

7

Dimensioning exercise

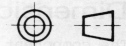

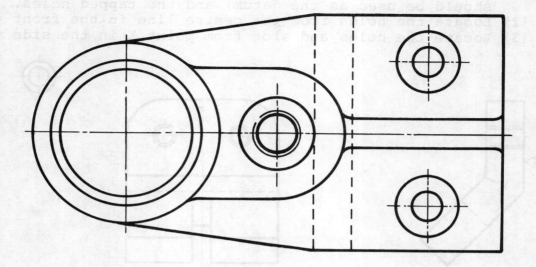

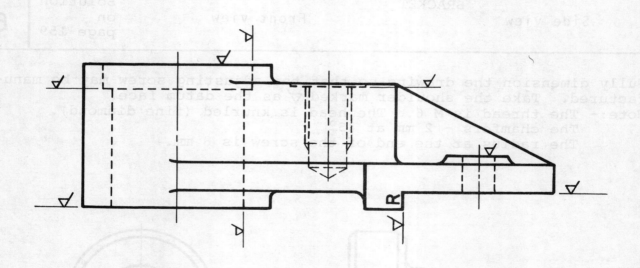

8

Fully dimension the orthographic views shown above to enable the bracket to be manufactured. Details of sizes, locations, screw threads, etc., may be taken from the pictorial view of the same component shown on the opposite page. (All dimensions millimetres) Both the 8 mm base holes and the 35 mm bore must be located from the machined face R. The tapped hole must be located from the 35 mm bore.

Note: This drawing can be fully and clearly dimensioned using two views only. The drawing of a more complicated component would possibly include a side view also to clarify both shape and dimensioning. The solution to this exercise is on page 160.

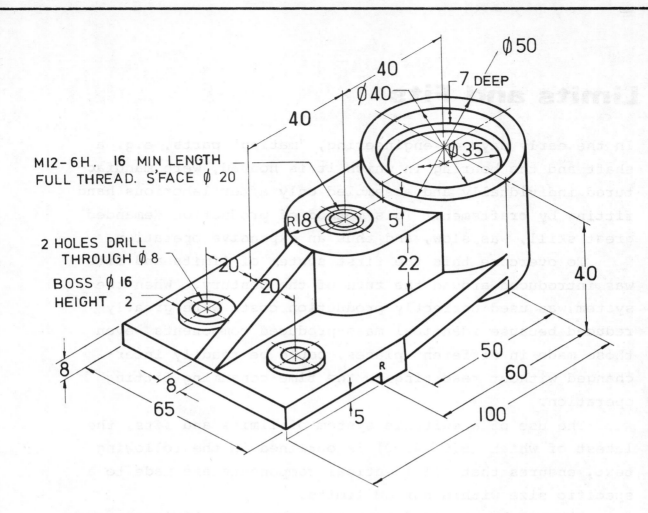

ø 50

7 DEEP

40

ø 40

40

ø 35

M12-6H. 16 MIN LENGTH
FULL THREAD. S'FACE ø 20

R18

5

40

2 HOLES DRILL
THROUGH ø 8

BOSS ø 16
HEIGHT 2

20

20

22

8

8

50

60

65

R

5

100

In this isometric view of the cast-iron
bracket all the corner fillets
(or radii) have been omitted
in order to simplify the
drawing.

Cast iron
Bracket
for
Vertical
Spindle.

Limits and Fits

In the early days of engineering, "mating" parts, e.g. a shaft and the bearing in which it is housed, were manufactured individually and assembled only after laborious hand fitting by craftsmen. This method of production demanded great skill, was slow, and thus an expensive operation.

To overcome this the first system of limits and fits was introduced around the turn of the century. When the system was used correctly production costs were greatly reduced because identical mass-produced components, even those made in different places, could be readily interchanged without resorting to the time-consuming fitting operation.

The use of a suitable system of limits and fits, the latest of which (B.S. 4500) is outlined in the following text, ensures that all identical components are made to a specific size within narrow limits.

It must be stressed that when assigning limits to the dimensions of a component it is necessary to consider not only its function but also the skill of the operator and the type and condition of the machines which are used to manufacture the component.

Limits should be applied to the dimensions of a component only if the resulting degree of accuracy is essential for the efficient functioning of the component. It is uneconomical to limit the accuracy of a dimension to ± 0.005 mm when ± 1 mm would prove satisfactory.

Limits and Fits for Holes and Shafts

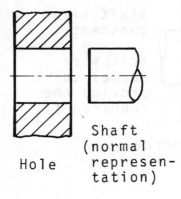

Hole   Shaft (normal representation)

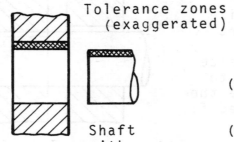

Tolerance zones (exaggerated)

Hole   Shaft with tolerance zones

A TOLERANCE on a hole or a shaft is a permissible degree or error. It depends on:

(1) Machining or cutting operation used.

(2) Quality of work.

(3) Type of fit between hole and shaft.

Effect of manufacturing process on degree of accuracy obtainable

Numerical value of tolerance for sample sizes

Tolerance number	Process (typical)	BS 1916 inch 0.001 in			BS 4500 metric 0.001 mm		
		0.119 to 0.237	1.969 to 3.150	4.725 to 7.087	3 mm to 6 mm	50 mm to 80 mm	120 mm to 180 mm
1	Slip blocks, reference gauges	0.06	0.08	0.16	1.0	2	3.5
2	High quality gauges	0.08	0.12	0.2	1.5	3	5
3	Good quality gauges	0.12	0.2	0.32	2.5	5	8
4	Gauges and precision fits	0.15	0.3	0.5	4	8	12
5	Ball bearings, fine grinding	0.2	0.5	0.7	5	13	18
6	Grinding, fine honing	0.3	0.7	1.0	8	19	25
7	High quality turning, broaching	0.5	1.2	1.6	12	30	40
8	Centre lathe turning and reaming	0.7	1.8	2.5	18	46	63
9	Horiz. and vert. boring	1.2	3.0	4.0	30	74	100
10	Milling, planing, extrusion	1.8	4.5	6.0	48	120	160
11	Drilling, rough turning, boring	3.0	7.0	10.0	75	190	250
12	Light press work, tube drawing	5.0	12.0	16.0	120	300	400
13	Press work, tube rolling	7.0	18.0	25.0	180	460	630
14	Die casting or moulding	12.0	30.0	40.0	300	750	1000
15	Stamping (approx)	18.0	45.0	60.0	480	1200	1600
16	Sand casting, flame cutting	30.0	70.0	100.0	750	1900	2500

Tolerance increases with the number

For more detailed lists refer to BS 1916 and BS 4500

Tolerance increases with size of component

The tolerance number is directly related to the manufacturing process and hence to the quality of work in terms of the degree of dimensional accuracy required. The permissible degree of error increases (*a*) as the process becomes less precise and (*b*) as the size of component increases.

Tolerance number should not be confused with the numbers which relate to surface roughness.

Refer to British Standards for further information.
     Use BS 1916 for inch sizes
     and BS 4500 for metric sizes.

Limit systems

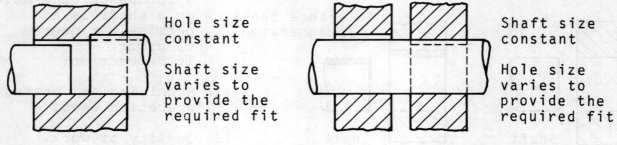

Hole size constant

Shaft size varies to provide the required fit

Clearance   Interference

HOLE BASIS

Shaft size constant

Hole size varies to provide the required fit

Clearance   Interference

SHAFT BASIS

BS 1916 and BS 4500 use the hole basis but there are occasions when it is necessary to design a fit on a shaft basis.

Basic size and disposition of tolerance

Whenever possible, holes are designed and machined to sizes obtained with standard tools and/or to preferred sizes in millimetres (see BS 4318).

The basic size of a hole is the dimension used if no tolerances were applied. Another name used is "nominal size".

Tolerances can be applied to holes and shafts in various ways.

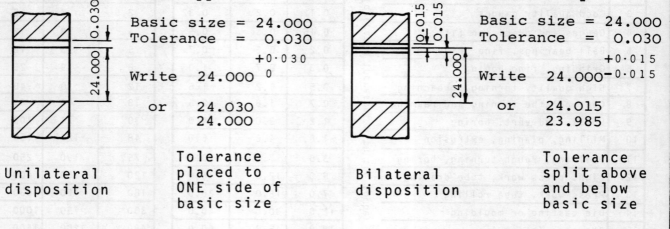

Basic size = 24.000
Tolerance = 0.030

Write 24.000 $^{+0.030}_{0}$

or   24.030
       24.000

Unilateral disposition

Tolerance placed to ONE side of basic size

Basic size = 24.000
Tolerance = 0.030

Write 24.000 $^{+0.015}_{-0.015}$

or   24.015
       23.985

Bilateral disposition

Tolerance split above and below basic size

BS 1916 and BS 4500 generally use the unilateral method of applying limits to a size.

The tolerances can be represented as shapes on a graph. The OX axis is the top side of the basic size, in this case, the hole.

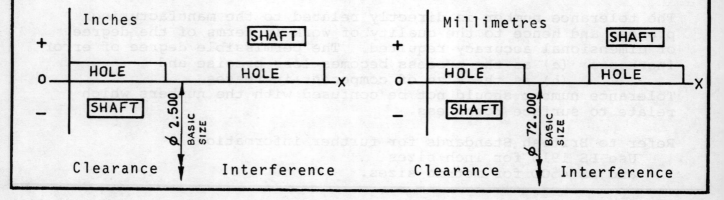

Inches

Millimetres

Clearance   Interference

Clearance   Interference

FIT: the relationship between hole (or internal feature) and shaft (or external feature)

ALLOWANCE - the prescribed (algebraic) difference between the low limit of size for the hole (or internal feature) and the high limit of size for the "mating" shaft (or external feature), i.e. maximum metal condition.

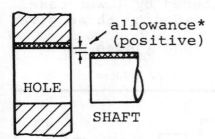

allowance* (positive)

HOLE

SHAFT

e.g. a simple bearing (exaggerated).

BEARING

The shaft is "loose" running or "easy" running or "slide".

Verbal descriptions of the fit may vary widely. Numerical differences are more important.

CLEARANCE FIT
Shaft is ALWAYS smaller than hole.

e.g. for φ2 inch (BS1916:53)

Hole sizes	2.0018
	2.0000
"Mating" shaft sizes	1.9988
	1.9970
Smallest hole	2.0000
Largest shaft	1.9988
ALLOWANCE	+ 0.0012

for φ24 mm (BS4500:69)

Hole sizes	24.033
	24.000
"Mating" shaft sizes	23.980
	23.959
Smallest hole	24.000
Largest shaft	23.980
ALLOWANCE	+ 0.020

Note! Allowance is ALWAYS positive (+) for a clearance fit.

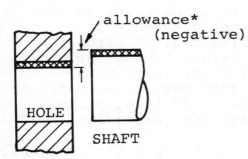

allowance* (negative)

HOLE

SHAFT

e.g. a gear wheel on a shaft (exaggerated)

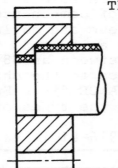

GEAR

The shaft is forced into the hole.

The fit can be described in such terms as:
"light press" or "press" or "heavy press" or "shrink"

Numerical differences are more important.

INTERFERENCE FIT
Shaft is ALWAYS larger than hole.

e.g. for φ2 inch (BS1916:53)

Hole sizes	(as above)
"Mating" shaft sizes	2.0032
	2.0020
Smallest hole	2.0000
Largest shaft	2.0032
ALLOWANCE	- 0.0032

for φ24 mm (BS4500:69)

Hole sizes	(as above)
"Mating" shaft sizes	24.048
	24.035
Smallest hole	24.000
Largest shaft	24.048
ALLOWANCE	- 0.048

As shaft sizes are subtracted from hole sizes, the allowance will always be negative (-) for an interference fit.
* The use of the word "allowance" has been omitted from BS4500. It has been used here solely to explain the difference between clearance fit and interference fit. It is still used, of course, in BS1916:53.

Both BS 1916 and BS 4500 recommend limits and fits for a wide range
of engineering processes and activities - even horology! For most
general purposes a carefully selected range of hole and shaft sizes
are given in each standard.

Holes are lettered by capitals, A,B,C, etc., the limits allocated
to the H-hole being recommended. The H-hole always has the basic
size of the hole as one size. The letter is followed by the
tolerance grade number e.g. H8. Shafts are lettered by lower case
letters, a,b,c, etc., and a range of fits is provided by those given
in the charts below.

BS 1916 φ1.182-1.575 in   BS 4500 φ30-40 mm

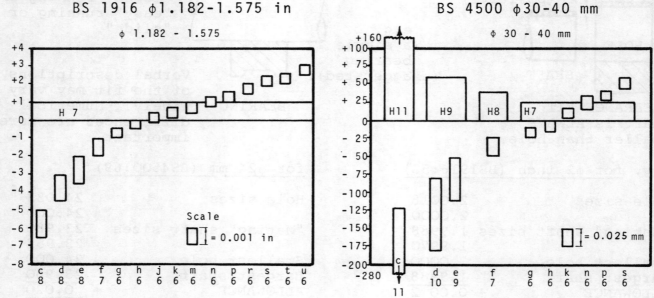

Tolerance grade number

If the tolerances on hole and shaft overlap it is possible in the
random selection of components to obtain either clearance or
interference. These are called TRANSITION fits e.g. H7 with k6,
and are described as "push", "easy keying", "tight keying" or
"drive". A fit between an H8 hole and an f7 shaft is written
H8-f7 and reference to the chart above will show that this must
always be a CLEARANCE fit. Further reference to BS 1916, parts
2 and 3 will show the types of fit recommended for particular
types of work.

Extract from BS 1916-Limit values. Extract from BS 4500-Limit values.

Hole H7 0.001 in	Size of component between		Shafts				
			e8 −	f7 −	h6 −	j6	s6 +
+ 0.5 0	in 0.12	in 0.24	− 0.8 − 1.5	− 0.4 − 0.9	0 − 0.3	+ 0.2 − 0.1	+ 1.0 + 0.7
+ 0.6 0	0.24	0.40	− 1.0 − 1.9	− 0.5 − 1.1	0 − 0.4	+ 0.3 − 0.1	+ 1.4 + 1.0
+ 0.7 0	0.40	0.71	− 1.2 − 2.2	− 0.6 − 1.3	0 − 0.4	+ 0.3 − 0.1	+ 1.6 + 1.2
+ 0.8 0	0.71	1.19	− 1.6 − 2.8	− 0.8 − 1.6	0 − 0.5	+ 0.3 − 0.2	+ 1.9 + 1.4
+ 1.0 0	1.19	1.97	− 2.0 − 3.6	− 1.0 − 2.0	0 − 0.2	+ 0.4 − 0.2	+ 2.4 + 1.8
+ 1.2	1.97	2.56	− 2.5 − 4.3	− 1.2 − 2.4	0 − 0.7	+ 0.4 − 0.3	+ 2.7 + 2.0
0	2.56	3.15					+ 2.9 + 2.2

	Hole H7 0.001 mm	Size of component between		Shafts			
				g6 −	k6 +	n6 +	s6 +
ES EI	+ 15 0	mm 6	mm 10	− 5 − 14	+ 10 + 1	+ 19 + 10	+ 32 + 23 es ei
ES EI	+ 18 0	10	18	− 6 − 17	+ 12 + 1	+ 23 + 12	+ 39 + 28 es ei
ES EI	+ 21 0	18	30	− 7 − 20	+ 15 + 2	+ 28 + 15	+ 48 + 35 es ei
ES EI	+ 25 0	30	40	− 9 − 25	+ 18 + 2	+ 33 + 17	+ 59 + 43 es ei
		40	50				
ES EI	+ 30 0	50	65	− 10 − 29	+ 21 + 2	+ 39 + 20	+ 72 + 53 es ei
		65	80				+ 78 + 59 es ei

Conventional method of illustrating terms

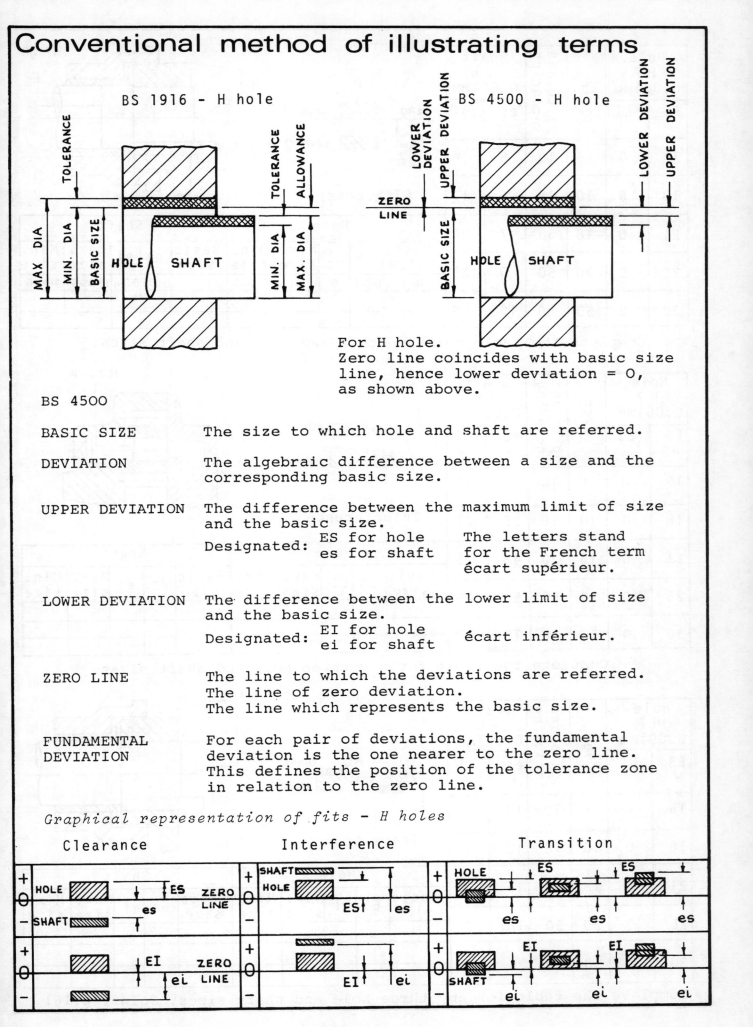

BS 1916 - H hole

BS 4500 - H hole

For H hole.
Zero line coincides with basic size
line, hence lower deviation = 0,
as shown above.

BS 4500

BASIC SIZE	The size to which hole and shaft are referred.
DEVIATION	The algebraic difference between a size and the corresponding basic size.
UPPER DEVIATION	The difference between the maximum limit of size and the basic size.

Designated: ES for hole / es for shaft — The letters stand for the French term écart supérieur.

LOWER DEVIATION   The difference between the lower limit of size and the basic size.

Designated: EI for hole / ei for shaft — écart inférieur.

ZERO LINE   The line to which the deviations are referred.
The line of zero deviation.
The line which represents the basic size.

FUNDAMENTAL DEVIATION   For each pair of deviations, the fundamental deviation is the one nearer to the zero line. This defines the position of the tolerance zone in relation to the zero line.

Graphical representation of fits - H holes

Clearance   Interference   Transition

Hole H 7 0.001mm		Over	Up to and including	Shaft g 6 0.001mm	
ES +	EI			es -	ei -
15	0	6	10	5	14
18	0	10	18	6	17
21	0	18	30	7	20
25	0	30	50	9	25
30	0	50	80	10	29

H7-g6

Fit

HOLE H 7
SHAFT g 6

Clearance

Hole					Shaft				
Basic size	ES	EI	Max. size	Min. size	Basic size	es	ei	Max. size	Min. size
10	0·015	0	10·015	10·000	10	0·005	0·014	9·995	9·986

Complete the table for any other two hole and shaft sizes.

Hole H 7 0.001mm		Over	Up to and including	Shaft s 6 0.001mm	
ES +	EI			es +	ei +
15	0	6	10	32	23
18	0	10	18	39	28
21	0	18	30	48	35
25	0	30	50	59	43
30	0	50	80	78	50

H7-s6

Fit

s6 SHAFT
H7 HOLE

Interference

Hole					Shaft				
Basic size	ES	EI	Max. size	Min. size	Basic size	es	ei	Max. size	Min. size

Complete the table for any three hole and shaft sizes.

Hole H 7 0.001mm		Over	Up to and including	Shaft k 6 0.001mm	
ES +	EI			es +	ei +
15	0	6	10	10	1
18	0	10	18	12	1
21	0	18	30	15	2
25	0	30	50	18	2
30	0	50	80	21	2

H7-k6

Fit

k6 SHAFT
H7 HOLE

Transition

Hole					Shaft				
Basic size	ES	EI	Max. size	Min. size	Basic size	es	ei	Max. size	Min. size

Complete the table for any three hole and shaft sizes. Solns. p.161.

ISO Metric Screw Threads (BS3643)

One of the most common items which needs to be clearly described on an engineering drawing is the screw thread which may be used for:

(1) Transmitting power, e.g. square thread in a vice or a lathe.
(2) Adjusting parts relative to each other (usually a Vee-thread).
(3) Fastening parts together, e.g. a nut and bolt (a Vee-thread).

The following section deals mainly with the ISO metric thread which is recognized internationally and, over the next few years, will be used in preference to the many varied types of threads which exist in industry today. In order to specify an ISO metric thread on a drawing it is necessary to understand the basic terminology of screw threads and how the fit between internal and external threads is derived.

The fit is based on the "effective" diameter, i.e. the diameter at which the pitch is measured.

Note: The helix and thread form are drawn simplified.

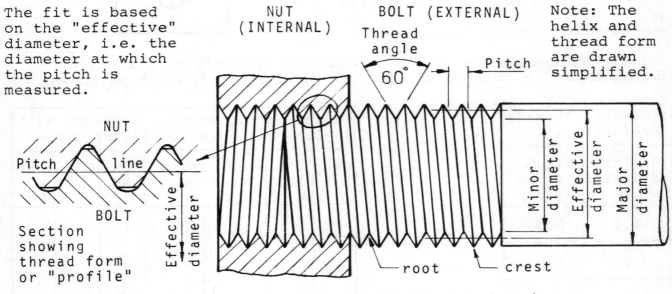

Class of fit

Although the thread profile is a Vee in section, the nuts are classed as "holes" and the bolts as "shafts" and tolerances are applied according to BS 4500 (Limits and Fits). They may be represented graphically like any hole and shaft combination.

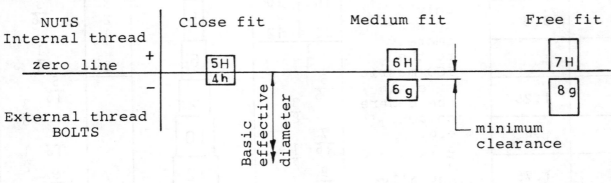

Fit designated as → 5H/4h   6H/6g   7H/8g

Coarse and fine pitches are available but it is expected that the coarse pitch screw threads will suffice for most applications and that the medium fit will be the most commonly used. This may be summarized as follows:

Class of fit	Bolts & screws	Nuts
medium	6g	6H

ISO Metric screw thread designation

In production engineering it is necessary to provide information about the size of thread and the tolerance allocated to it. On a drawing, the thread would be "dimensioned" or designated by stating the information as shown in the following examples:

M 10 × 1.5 - 6H   [NUT - capital H for hole or internal thread.]

M 10 × 1.5 - 6g   [BOLT - small g for shaft or external thread.]

- Thread tolerance grade number
- Pitch in millimetres
- Nominal size in millimetres
- Symbol for ISO metric.

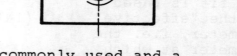

—M10-6H

The ISO metric coarse is the thread most commonly used and a designation for this would omit the pitch, e.g. M 10-6H as shown.

Isometric threads Preferred sizes	
Nominal diameter mm	Pitch coarse mm
1.6	0.35
2	0.4
2.5	0.45
3	0.5
4	0.7
5	0.8
6	1
8	1.25
10	1.5
12	1.75
16	2
20	2.5
24	3

Note

In the table on the right, the inch and number sizes relate only approximately to the ISO metric sizes in the centre column. e.g. $\frac{5}{8}$ in actually equals 15.875 mm

13 ISO metric threads compared with
74 Imperial inch and number sizes

UNC No. DIA	UNF No.	ISO metric DIA mm	BA No.	BSW DIA	BSF DIA
	0	1·6	10		
1	1	2	9		
2	2		8		
3	3	2·5	7	$\frac{1}{8}$	
4	4	3	6	$\frac{5}{32}$	
5	5		5		
6	6	4	4		
8	8		3		
10	10	5	2		$\frac{3}{16}$
12	12		1		
$\frac{1}{4}$		6	0		$\frac{1}{4}$
$\frac{5}{16}$	$\frac{3}{8}$	8		$\frac{5}{16}$	$\frac{3}{8}$
$\frac{7}{16}$	$\frac{1}{2}$	10		$\frac{7}{16}$	$\frac{1}{2}$
$\frac{9}{16}$	$\frac{5}{8}$	12		$\frac{9}{16}$	$\frac{5}{8}$
$\frac{3}{4}$	$\frac{7}{8}$	16		$\frac{3}{4}$	$\frac{7}{8}$
1		20		1	
		24			

Screw threads, nuts and bolts (ISO metric series)

A rapid, easy-to-remember method of drawing screw threads, nuts and bolts is essential if templates are not available. The drawing should look right and slight discrepancies in size are not important. The most used ISO metric bolts (up to ϕ24 mm) are:

	M6	M8	M10	M12	M14	M16	M20	M22	M24
Major diameter mm	6·00	8·00	10·00	12·00	14·00	16·00	20·00	22·00	24·00
Minor diameter	4·77	6·47	8·16	9·85	11·55	13·55	16·93	18·93	20·32
Minor diameter (approx.)	4·8	6·5	8·2	9·8	11·6	13·6	17·0	19·0	20·4

<u>Example of application</u> M 16 Stud

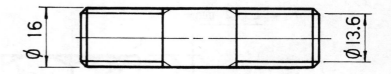

Nut and bolt head sizes across the flats (spanner sizes) in mm.

	6	8	10	12	14	16	20	22	24
Major diam. D. Distance across the flats	10	13	17	19	22	24	30	32	36
$D \times 1·5$						24	30	33	36
$(D \times 1·5) + 1$ mm	10	13	16	19	22				
Thickness of nut	5·0	6·5	8·0	10·0	11·0	13·0	16·0	18·0	19·0
Height of head	4·0	5·5	7·0	8·0	9·0	10·0	13·0	14·0	15·0

Approximate method: Thickness of nut = $D \times 0·8$ e.g. M10
Height of head   = $D \times 0·7$ e.g. M10

Example of application

M 16 Nut

Nuts are usually made from hexagonal stock bar and a 30° chamfer is machined on at least one surface.
In the plan view this appears as a circle, the diameter of which is equal to the distance across the flats.
The chamfer is drawn in the front view but <u>not</u> in the side view.

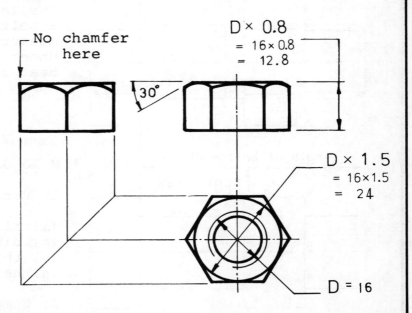

Nuts and bolts exercises

The drawing below is of an M 24 nut showing how the radii in the front view and side view may be constructed. The major diameter of the bolt is the size D for calculating the distance across the flats and the thickness of the nut.

Major diameter D = 24 mm
Minor diameter   = 20·4 mm approx.(See p.93).

$D \times 1·5 = 24 \times 1·5 = 36$ mm A/F

and R = 18 mm – this value of R is used
                    to find the centres for
                    radii (1), (2) and (3)

$D \times 0·8 = 24 \times 0·8 = 19·2$ mm

$T = 19·2$ mm

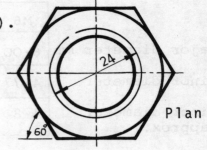

Plan

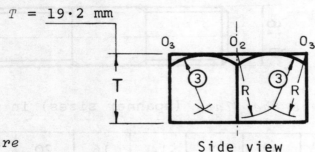

Side view

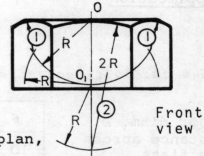

Front view

Procedure

(1) Draw major and minor diameter circles in plan, according to BS convention (See page 38).
(2) Set out circle in plan with radius = 18 mm, i.e. distance A/F.
(3) Construct hexagon around this circle using 60° set-square.
(4) Set out thickness of nut = 19·2 mm and project the hexagon to the front and side views.
(5) In the front view the centres for the radii marked (1) and (2) are obtained by setting out the radius R = 18 mm from O and O_1 as shown.
(6) In the side view the centres for the radii marked (3) are obtained by setting out the radius R from O_2 and O_3 as shown.

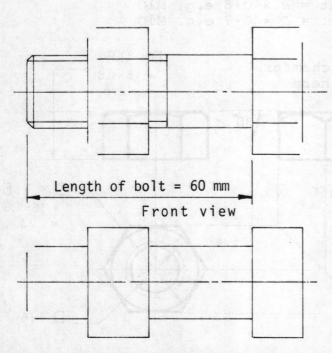

Length of bolt = 60 mm

Front view

Note:

The length of a bolt is measured from under the head as shown.

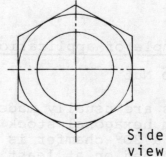

Side view

Exercise

M 20 Bolt with nut.

1. Verify that

   (a) minor diameter   = 17·0 mm
   (b) distance across flats = 30·0 mm
   (c) thickness of nut   = 16·0 mm
   (d) height of bolt head = 14·0 mm
                        (approx.)

2. Complete the drawing.

Assembly Drawings

The purpose of an assembly drawing is to provide visual information about the way in which parts of a machine or structure fit together. There are several types of assembly drawings and the differences in presentation depend on the uses for which they are intended. They are:

(1) LAYOUT ASSEMBLIES - in which the designer places together all the various parts in order to establish overall sizes, distances, etc., and, as a result, the feasibility of his design.

(2) OUTLINE ASSEMBLIES - these give general information about a machine or a group of components, for example, main sizes and centre distances which would show how the unit would be installed. This type of assembly is often used in catalogues giving details of the range of units offered for sale.

(3) GENERAL ASSEMBLIES or ARRANGEMENT DRAWINGS - show clearly how components fit together and, more important, how the assembled unit functions. Outside views, sectional and part-sectional views may be used but dimensions are rarely needed. The various parts may be labelled by ballooning and a parts list would complete the drawing.

(4) SUB-ASSEMBLIES - are drawings which show only one unit of a multi-unit component. On more complicated or multiple part components it may first be necessary to arrange parts into sub-assemblies which are then built up into the main assembly.

(5) SECTIONED ASSEMBLIES - a simple assembly may be drawn without the need for sectional views and be clearly understood. On more complex assembly drawings, however, too many hidden detail lines tend to confuse and a sectional view of the assembled parts conveys the information more clearly.

Note An assembly drawing should not be overloaded with detail. The correct place to fully dimension and describe a single part is on a detail drawing.

The drawings shown on the next page are typical of the use of a general assembly to give a general picture of a valve and a sub-assembly to show how the spindle unit is built up. A detail drawing of one of the parts is added to emphasise that the physical assembly of the valve depends on the manufacture of a large number of individual items.

The exercises which follow are of relatively simple assemblies which could be drawn with reasonable clarity without the use of sections but the resulting increase in clarity will show the advantage to be gained by their use.

The solution to the example on page 98 indicates the way in which the exercises should be carried out.

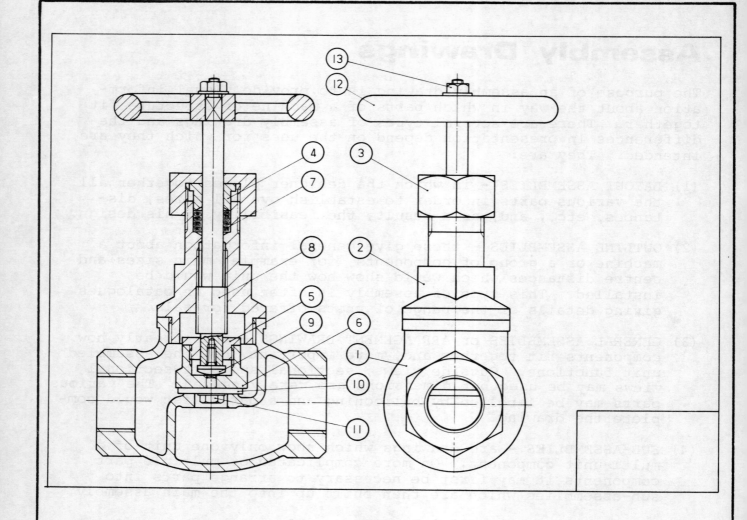

General assembly (or General arrangement)

The above drawing is a general assembly of a screwdown valve
used to control the flow of water.

The purpose of this sectional view is to label the various parts
of the valve and to show how they fit together. A parts list
which matches the ballooned letters is necessary and this may be
either on the assembly drawing or on a separate document.

The general assembly is not dimensioned in detail although the
sizes of the inlet and outlet and a few overall sizes may be
shown as the drawing may be typical of a range of valves of
differing sizes.

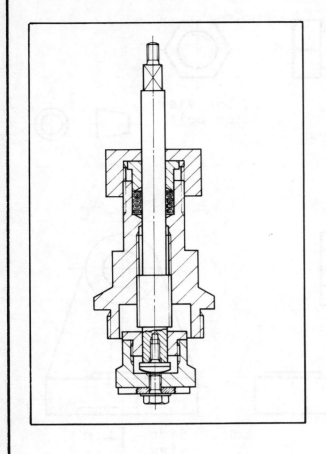

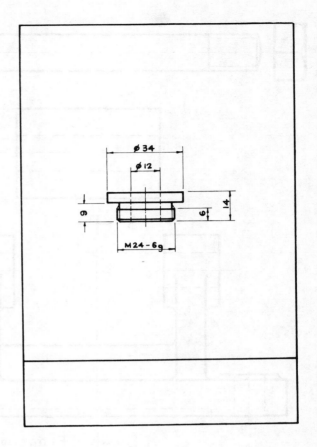

SUB-ASSEMBLY

The above sub-assembly
drawing shows how several parts
are assembled together before
being fitted into the main body
of the valve. This type of
drawing is particularly useful
if information more detailed
than that shown on the general
assembly is required, for
example, by the person
assembling the valve or the
maintenance worker.

DETAIL DRAWING

The detail drawing shown above
is of one of the numerous components which make up the complete valve and in this case is
a single-part drawing.

All the dimensions and processes
required to manufacture the part
must be shown on the drawing.

It is probable that the machin-
ist who makes this item will
take no part in assembling the
valve and will therefore use
only this detail drawing.

Sectioned assembly - Typical exercise

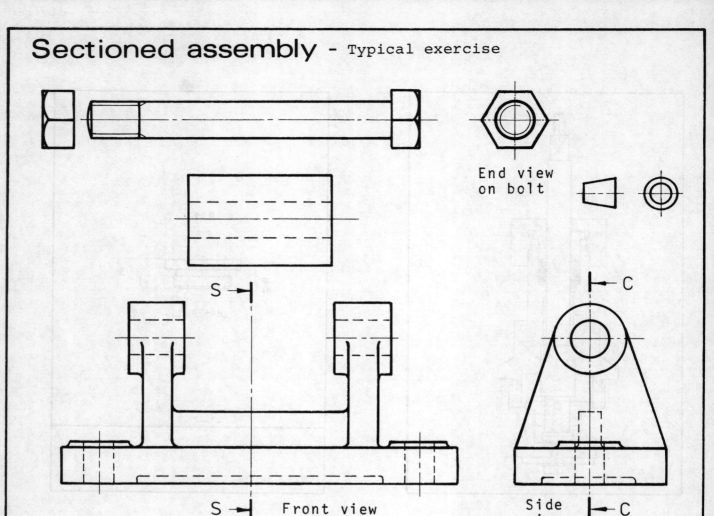

End view
on bolt

S

S — Front view

C

Side view — C

Exercise   Build up a sectioned assembly drawing of the component
parts looking on cutting plane CC to obtain a sectional
front view and cutting plane SS to obtain a sectional
end view, by tracing over and correctly positioning
each part.

Solution   Note that most of the outlines are unaltered.

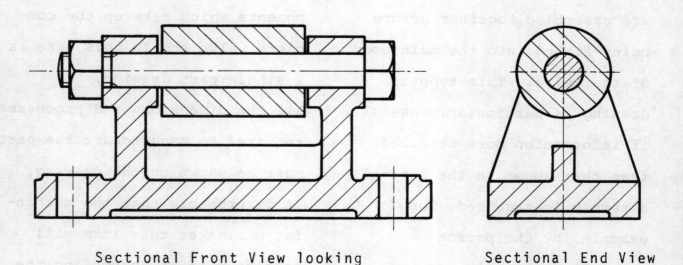

Sectional Front View looking
on cutting plane CC

Sectional End View
looking on
cutting plane SS.

Note The bolt and rib are <u>not</u>
sectioned in the front view, but <u>are</u> sectioned in the end view.

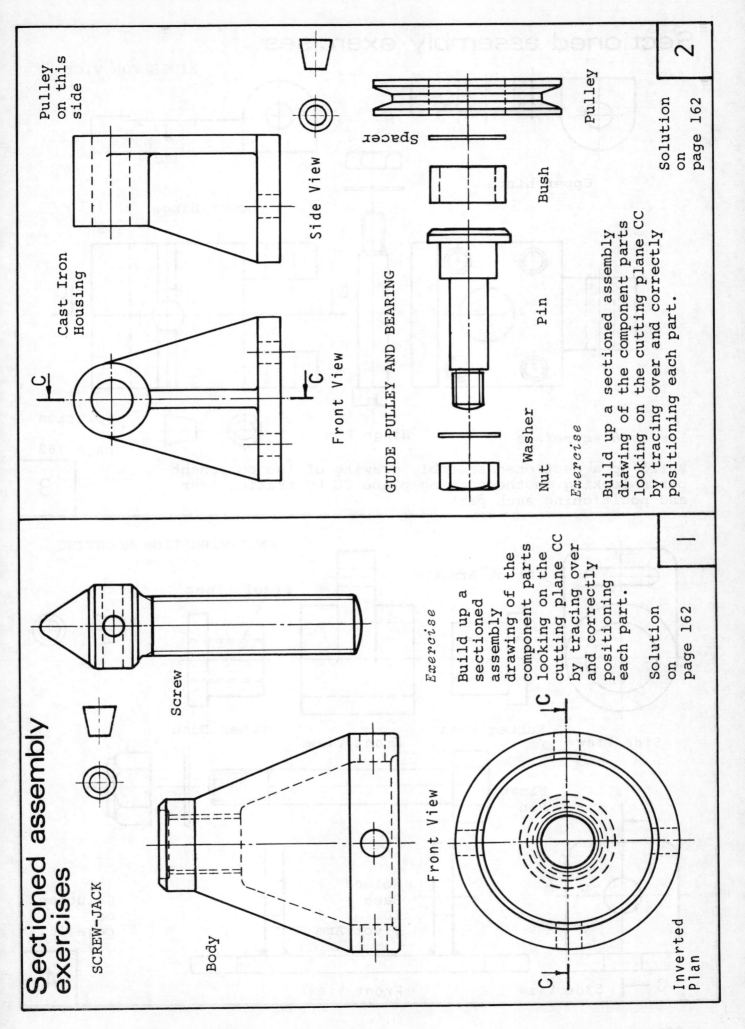

Sectioned assembly exercises

SCREW-JACK

Screw

Body

Front View

Inverted Plan

C

Exercise

Build up a sectioned assembly drawing of the component parts looking on the cutting plane CC by tracing over and correctly positioning each part.

Solution on page 162

1

Pulley on this side

Cast Iron Housing

Side View

Front View

C

C

GUIDE PULLEY AND BEARING

Spacer

Bush

Pin

Washer

Nut

Pulley

Exercise

Build up a sectioned assembly drawing of the component parts looking on the cutting plane CC by tracing over and correctly positioning each part.

Solution on page 162

2

99

Sectioned assembly exercises

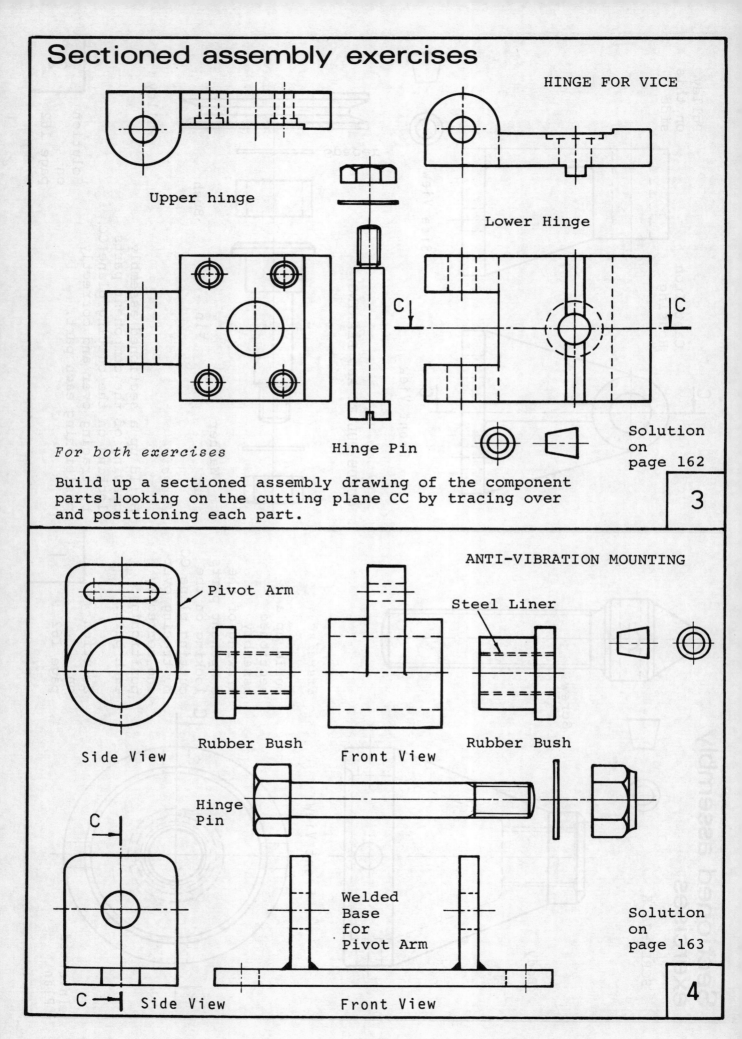

HINGE FOR VICE

Upper hinge

Lower Hinge

Hinge Pin

C

C

Solution on page 162

For both exercises

Build up a sectioned assembly drawing of the component parts looking on the cutting plane CC by tracing over and positioning each part.

3

ANTI-VIBRATION MOUNTING

Pivot Arm

Steel Liner

Side View

Rubber Bush

Front View

Rubber Bush

Hinge Pin

Welded Base for Pivot Arm

Solution on page 163

C

Side View

Front View

4

Sectioned assembly exercise

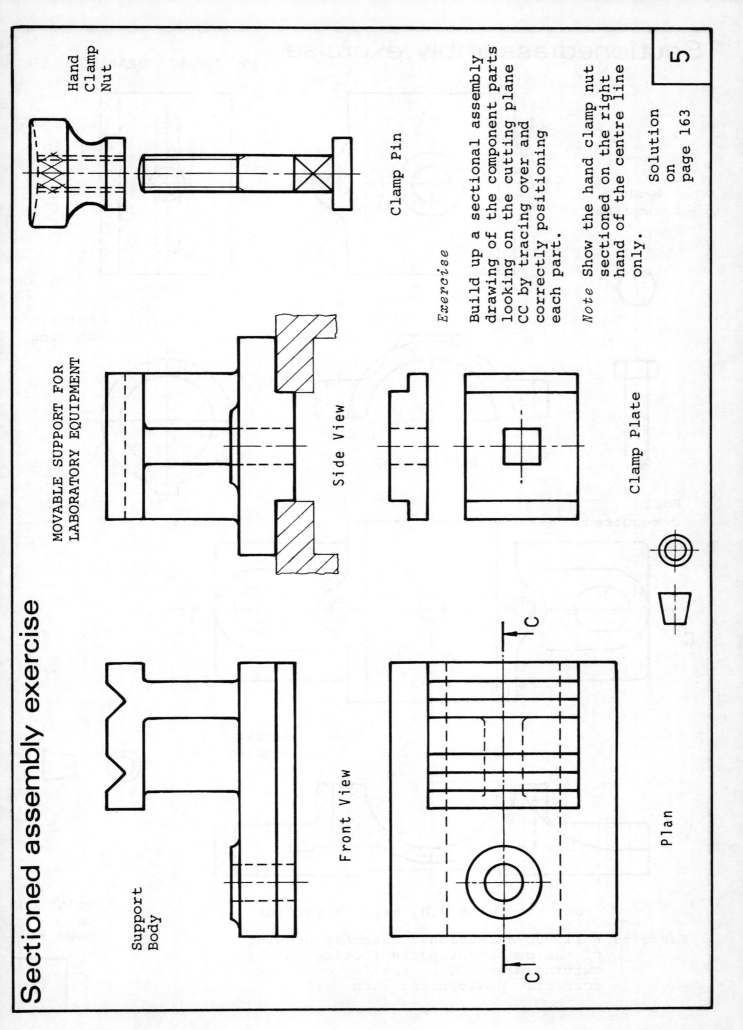

Hand
Clamp
Nut

Clamp Pin

MOVABLE SUPPORT FOR
LABORATORY EQUIPMENT

Side View

Clamp Plate

Support
Body

Front View

Plan

Exercise

Build up a sectional assembly
drawing of the component parts
looking on the cutting plane
CC by tracing over and
correctly positioning
each part.

Note Show the hand clamp nut
sectioned on the right
hand of the centre line
only.

Solution
on
page 163

5

Sectioned assembly exercise

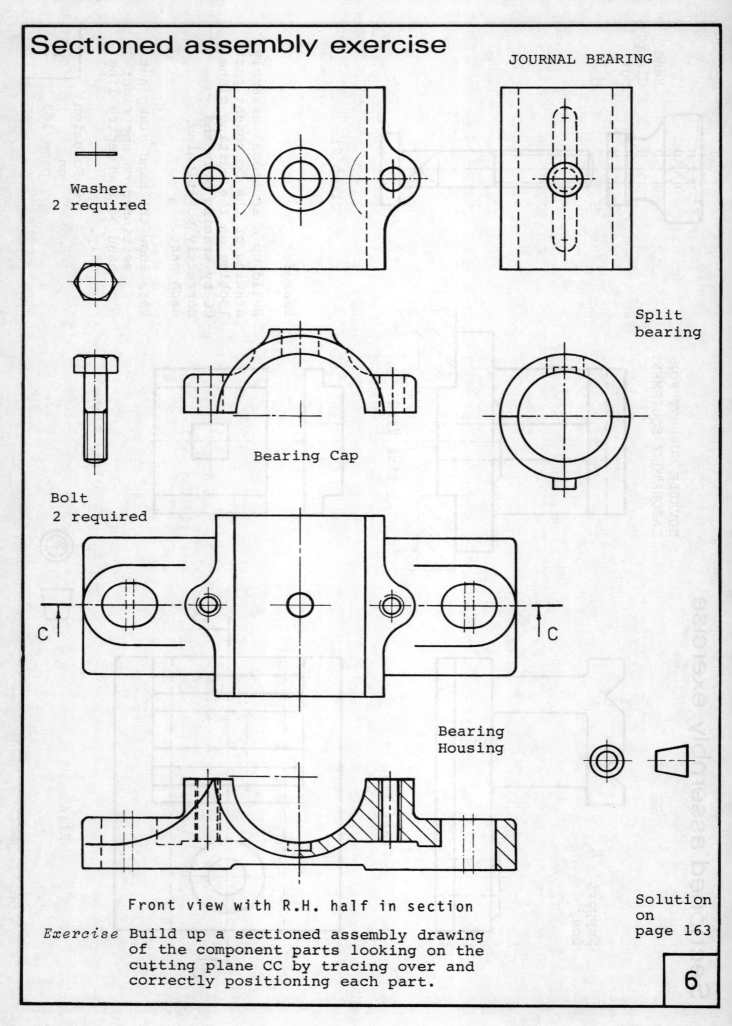

Washer
2 required

Bolt
2 required

Bearing Cap

Split
bearing

Bearing
Housing

C ↑   ↑ C

Front view with R.H. half in section

Solution
on
page 163

Exercise Build up a sectioned assembly drawing
of the component parts looking on the
cutting plane CC by tracing over and
correctly positioning each part.

6

Sectioned assembly exercise

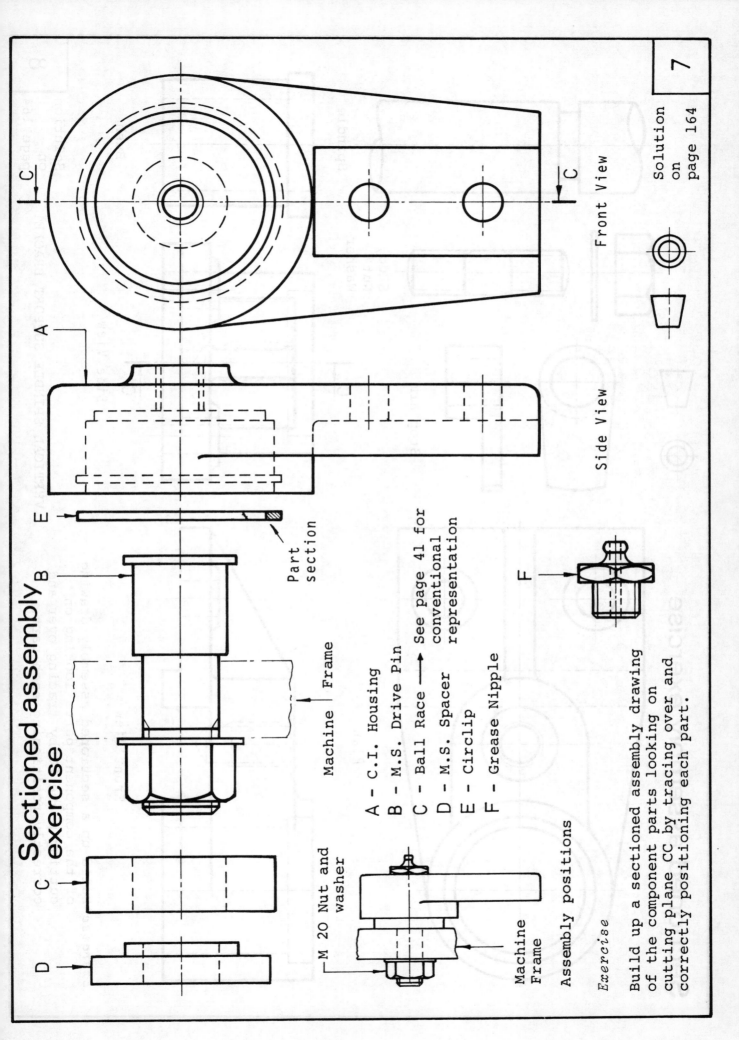

C ⟶ | C

Front View

Side View

A

E ⟶

Part section

B

Machine | Frame

C ⟶

D ⟶

M 20 Nut and washer

Machine Frame

Assembly positions

A – C.I. Housing
B – M.S. Drive Pin
C – Ball Race ⟶ See page 41 for conventional representation
D – M.S. Spacer
E – Circlip
F – Grease Nipple

F ⟶

Exercise

Build up a sectioned assembly drawing of the component parts looking on cutting plane CC by tracing over and correctly positioning each part.

Solution on page 164

7

103

Sectioned assembly exercise

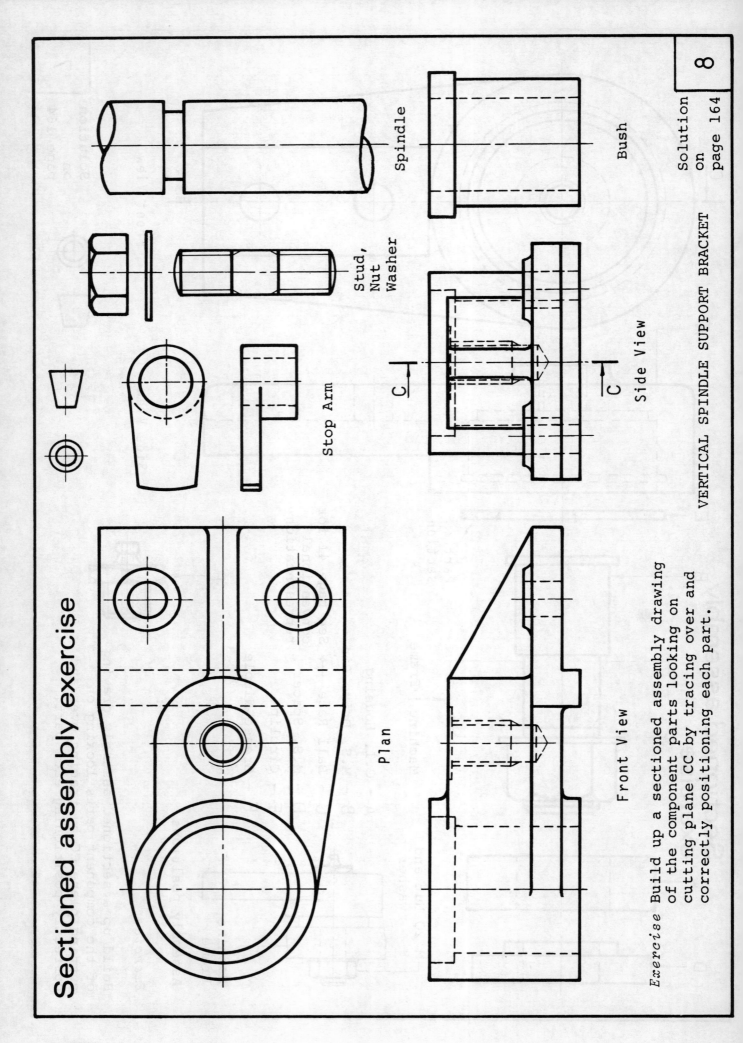

Spindle

Bush

Stud,
Nut
Washer

Stop Arm

C

C

Side View

Plan

Front View

Exercise Build up a sectioned assembly drawing of the component parts looking on cutting plane CC by tracing over and correctly positioning each part.

VERTICAL SPINDLE SUPPORT BRACKET

Solution
on
page 164

8

Developments

Many objects in everyday use are made from thin materials such as cardboard, plastic, aluminium, copper, brass and steel. Metals not more than about 3 mm thick are referred to as SHEET METALS.

The objects made from these materials are developed from the flat sheet. A pattern is drawn on the sheet which is then cut to the outline of the pattern. The material is bent, folded or rolled into the required shape, these processes giving the object stiffness and strength for a comparatively small weight of material.

A pattern which is used to mark out further patterns is called a TEMPLATE.

There are four basic shapes in sheet metal development from which a wide variety of work is fashioned, including hoppers, bins, chutes, vessels and ducts for ventilation and dust-extraction.

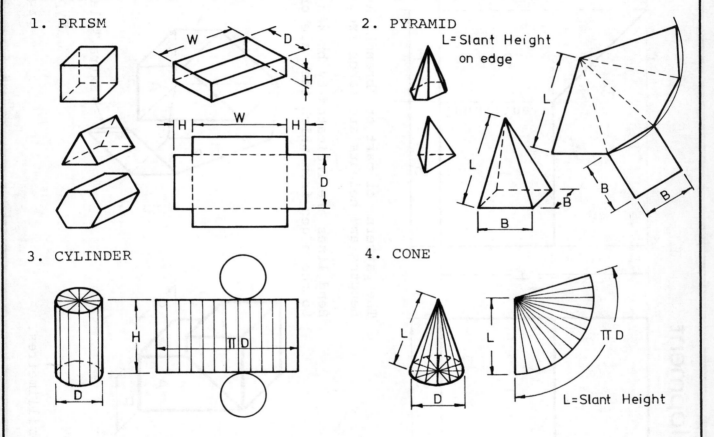

1. PRISM

2. PYRAMID
L=Slant Height on edge

3. CYLINDER

4. CONE
L=Slant Height

Simple shapes based on prisms, cylinders, pyramids and cones may be developed by three commonly used methods of construction. They are

1. PARALLEL LINE DEVELOPMENT: for prisms and cylinders mainly.

2. RADIAL LINE DEVELOPMENT: for pyramids and cones.

3. TRIANGULATION: for transition pieces which are used to join different shapes.

A pattern is usually folded so that the lines indicating its shape appear on the inside of the component. In practice, allowances may have to be made for extra material required for joints, stiffened edges, bends and seams.

No such allowances have been made in the exercises in the following pages so that basic principles can be emphasized.

Parallel line development

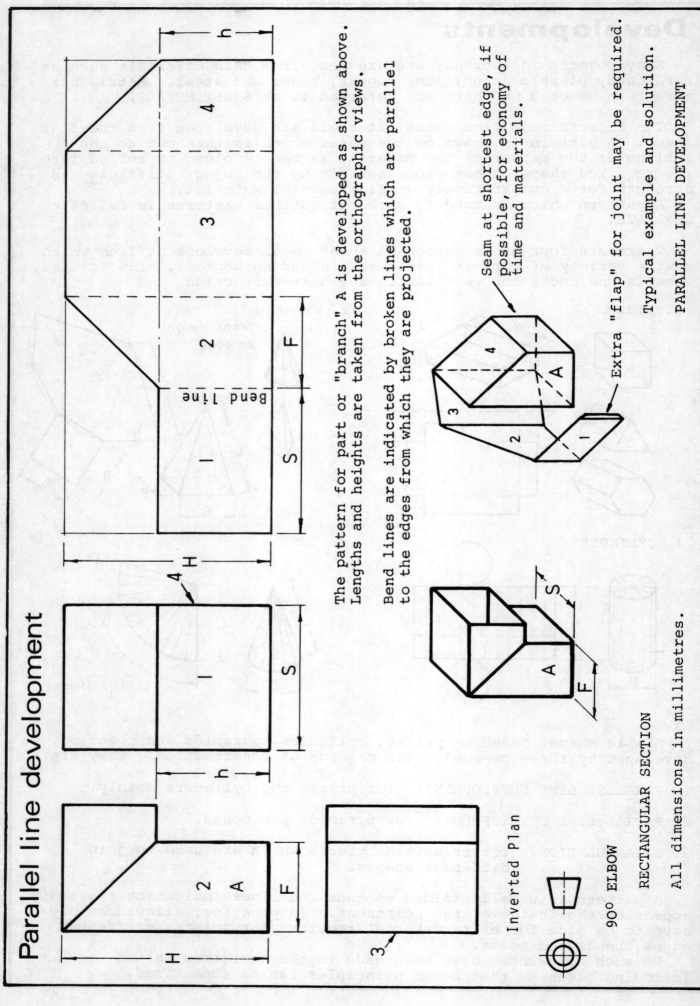

The pattern for part or "branch" A is developed as shown above. Lengths and heights are taken from the orthographic views.

Bend lines are indicated by broken lines which are parallel to the edges from which they are projected.

Bend line

Inverted Plan

90° ELBOW

RECTANGULAR SECTION

All dimensions in millimetres.

Seam at shortest edge, if possible, for economy of time and materials.

Extra "flap" for joint may be required.

Typical example and solution.

PARALLEL LINE DEVELOPMENT

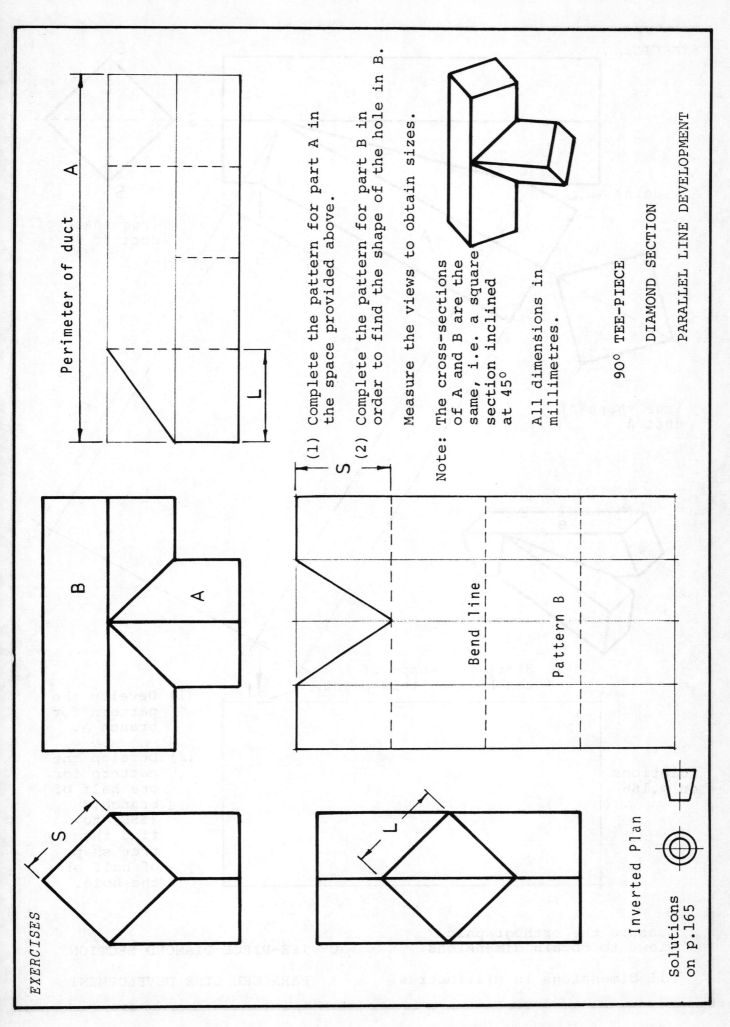

EXERCISES

Perimeter of duct A

L

S

B

A

S

L

Bend line

Pattern B

(1) Complete the pattern for part A in the space provided above.

(2) Complete the pattern for part B in order to find the shape of the hole in B.

Measure the views to obtain sizes.

Note: The cross-sections of A and B are the same, i.e. a square section inclined at 45°.

All dimensions in millimetres.

90° TEE-PIECE

DIAMOND SECTION

PARALLEL LINE DEVELOPMENT

Inverted Plan

Solutions on p.165

107

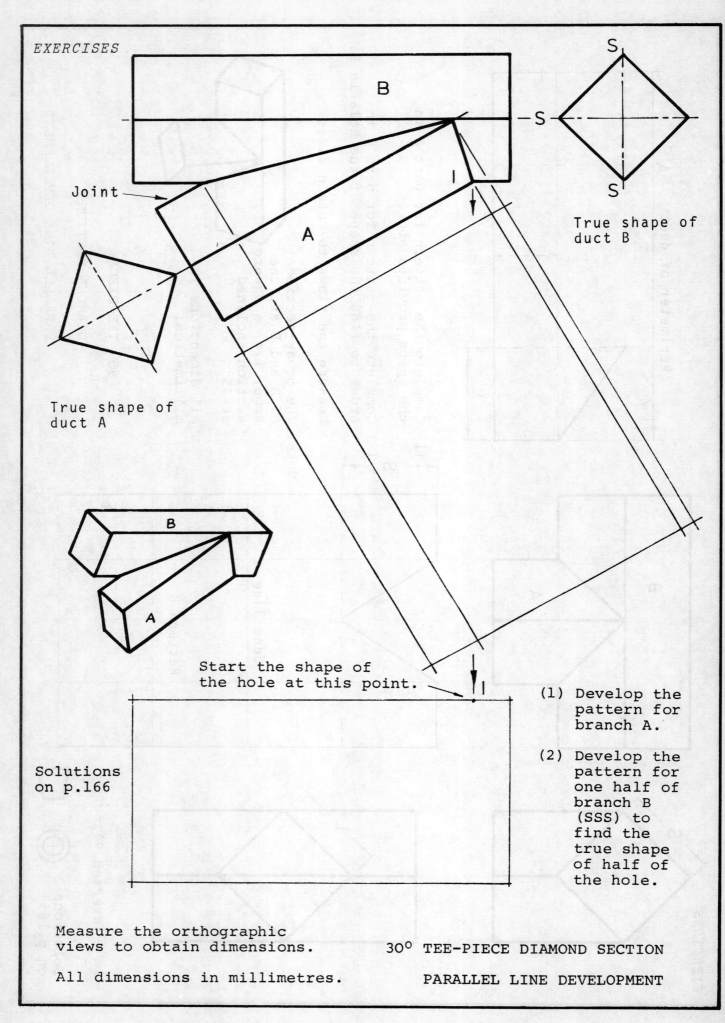

B

S
S
S

True shape of
duct B

Joint

A

True shape of
duct A

B

A

Start the shape of
the hole at this point.

(1) Develop the
pattern for
branch A.

(2) Develop the
pattern for
one half of
branch B
(SSS) to
find the
true shape
of half of
the hole.

Solutions
on p.166

Measure the orthographic
views to obtain dimensions.

All dimensions in millimetres.

30° TEE-PIECE DIAMOND SECTION

PARALLEL LINE DEVELOPMENT

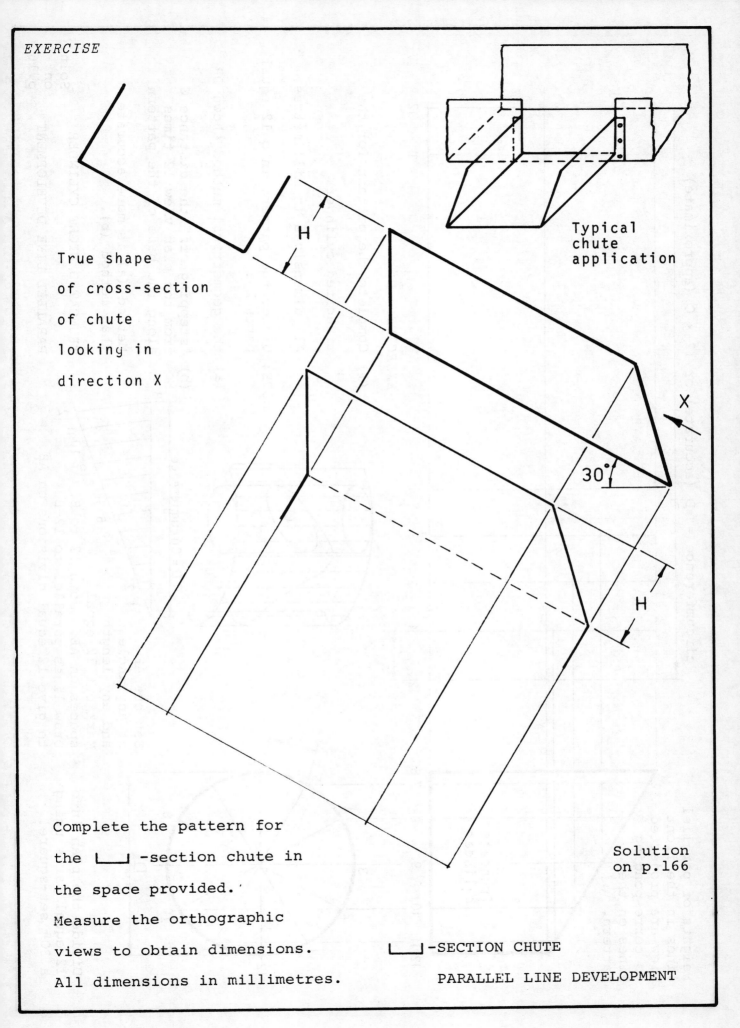

EXERCISE

Typical
chute
application

True shape
of cross-section
of chute
looking in
direction X

H

X

30°

H

Complete the pattern for

the ⌐⌐ -section chute in

the space provided.

Measure the orthographic

views to obtain dimensions.

All dimensions in millimetres.

Solution
on p.166

⌐⌐ -SECTION CHUTE

PARALLEL LINE DEVELOPMENT

109

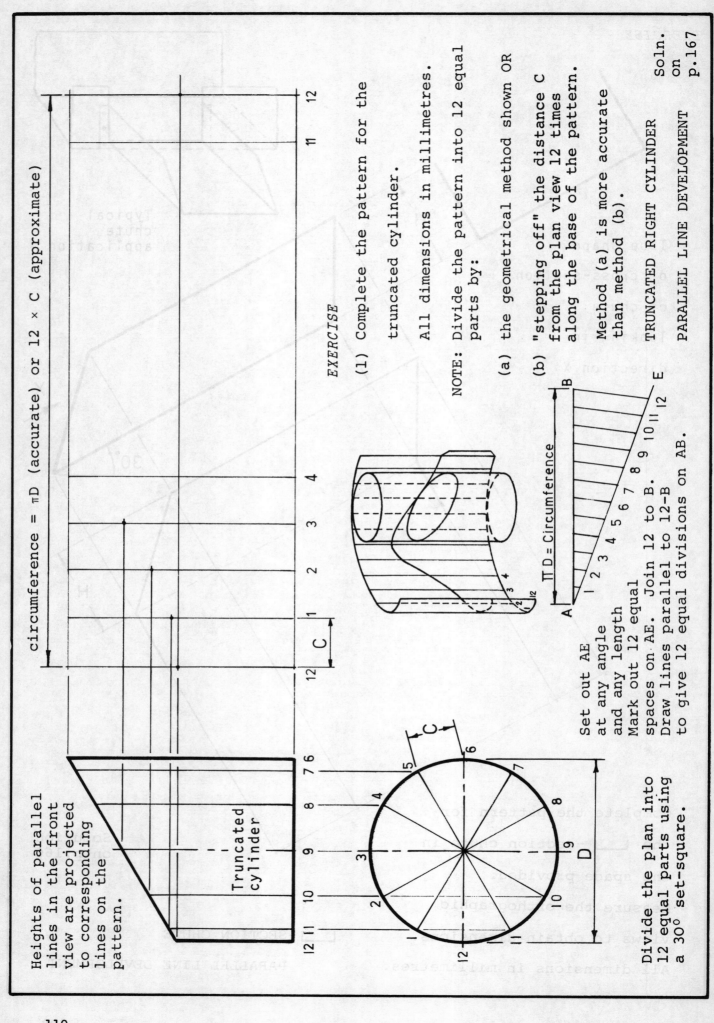

circumference = πD (accurate) or 12 × C (approximate)

Heights of parallel lines in the front view are projected to corresponding lines on the pattern.

Truncated cylinder

C

12 1 2 3 4 ... 11 12

12 11 10 9 8 7 6 5 4

C

D

Divide the plan into 12 equal parts using a 30° set-square.

EXERCISE

(1) Complete the pattern for the truncated cylinder.

All dimensions in millimetres.

NOTE: Divide the pattern into 12 equal parts by:

(a) the geometrical method shown OR

(b) "stepping off" the distance C from the plan view 12 times along the base of the pattern.

Method (a) is more accurate than method (b).

Set out AE at any angle and any length. Mark out 12 equal spaces on AE. Join 12 to B. Draw lines parallel to 12-B to give 12 equal divisions on AB.

A 1 2 3 4 5 6 7 8 9 10 11 12 E

B

πD = Circumference

TRUNCATED RIGHT CYLINDER   Soln. on p.167

PARALLEL LINE DEVELOPMENT

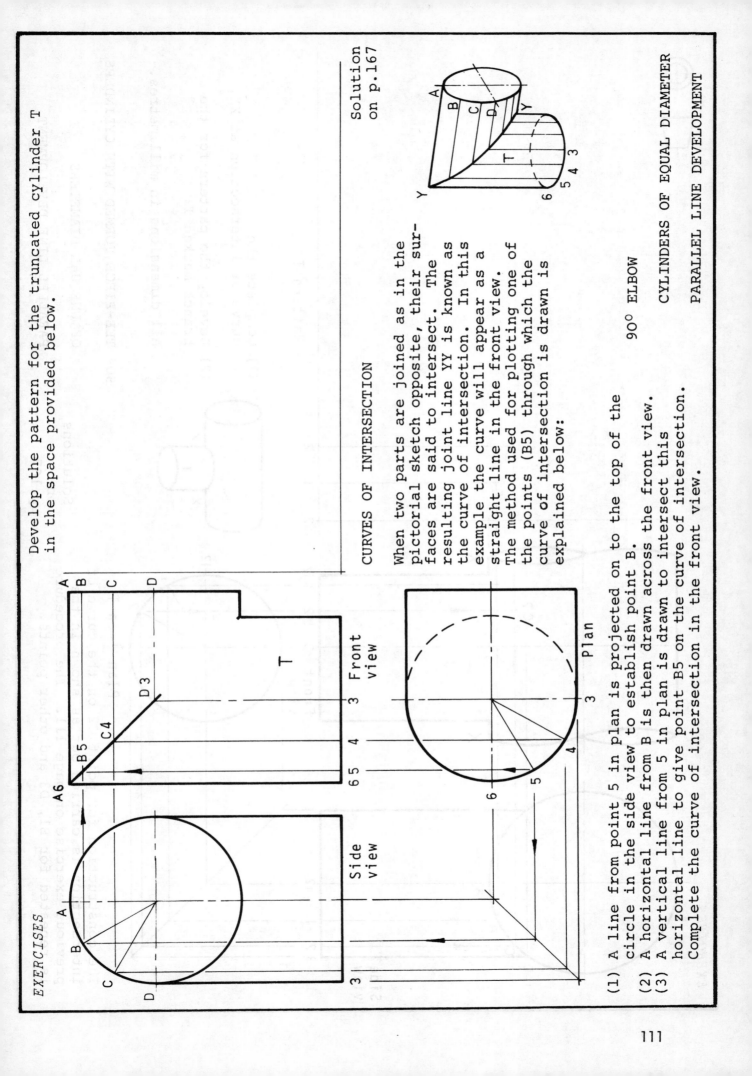

EXERCISES

Develop the pattern for the truncated cylinder T in the space provided below.

Front view

Side view

T

A B C D

3 4 6 5

A6 B5 C4 D3

Plan

Solution on p.167

CURVES OF INTERSECTION

When two parts are joined as in the pictorial sketch opposite, their surfaces are said to intersect. The resulting joint line YY is known as the curve of intersection. In this example the curve will appear as a straight line in the front view. The method used for plotting one of the points (B5) through which the curve of intersection is drawn is explained below:

A B C D Y

T Y

5 4 3 6

90° ELBOW

(1) A line from point 5 in plan is projected on to the top of the circle in the side view to establish point B.

(2) A horizontal line from B is then drawn across the front view. A vertical line from 5 in plan is drawn to intersect this horizontal line to give point B5 on the curve of intersection. Complete the curve of intersection in the front view.

CYLINDERS OF EQUAL DIAMETER
PARALLEL LINE DEVELOPMENT

111

EXERCISES

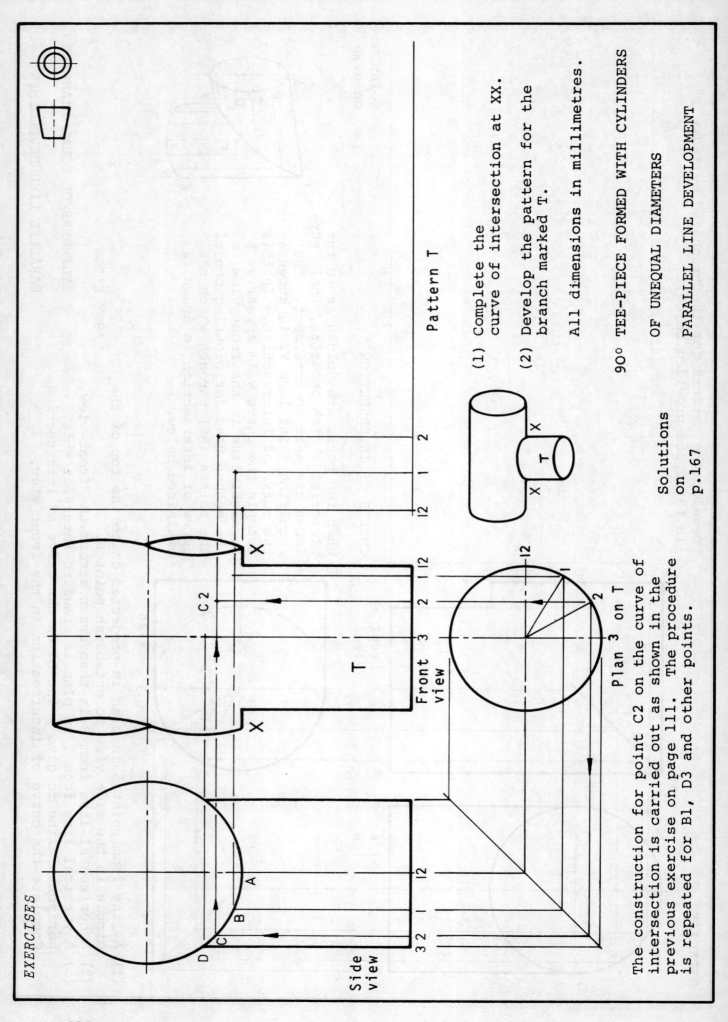

Side view

Front view

T

Plan 3 on T

Pattern T

The construction for point C2 on the curve of intersection is carried out as shown in the previous exercise on page 111. The procedure is repeated for B1, D3 and other points.

(1) Complete the curve of intersection at XX.

(2) Develop the pattern for the branch marked T.

All dimensions in millimetres.

90° TEE-PIECE FORMED WITH CYLINDERS

OF UNEQUAL DIAMETERS

PARALLEL LINE DEVELOPMENT

Solutions
on
p.167

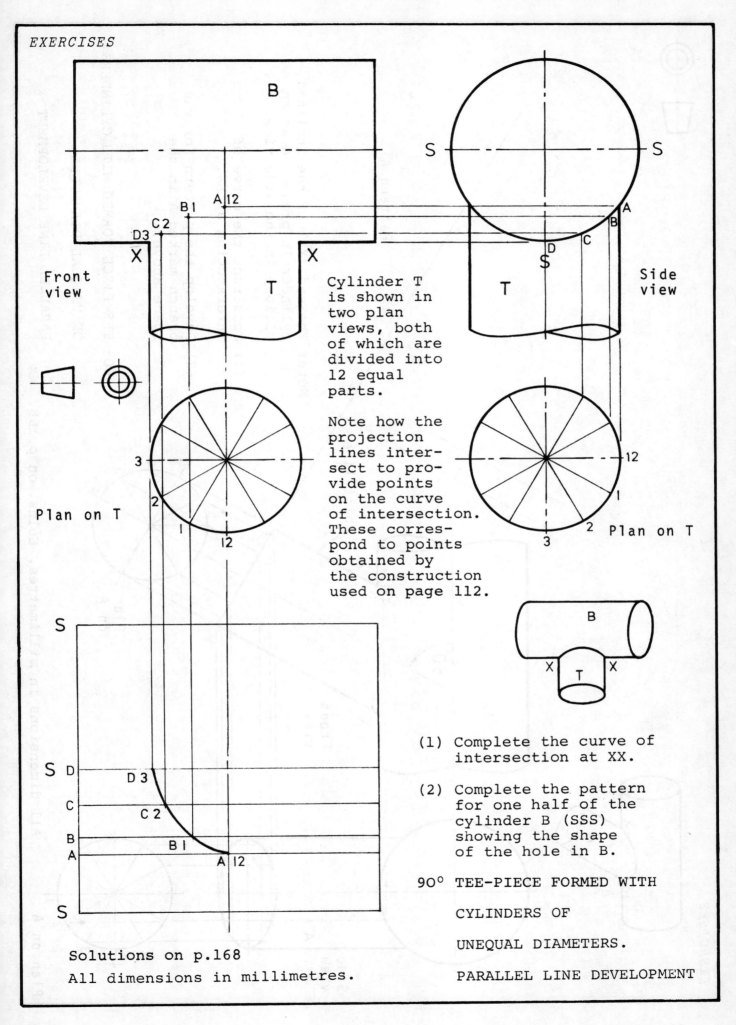

B

S ——————— S

A 12
B 1   A
C 2   B
D 3   D   C

X   X   S

Front
view   Side
view

T   T

Cylinder T
is shown in
two plan
views, both
of which are
divided into
12 equal
parts.

Plan on T

3   12

2   1

1   2
12   3   Plan on T

Note how the
projection
lines inter-
sect to pro-
vide points
on the curve
of intersection.
These corres-
pond to points
obtained by
the construction
used on page 112.

S

B

X   T   X

S D
D 3
C
C 2
B
B 1
A
A 12

S

(1) Complete the curve of
    intersection at XX.

(2) Complete the pattern
    for one half of the
    cylinder B (SSS)
    showing the shape
    of the hole in B.

90° TEE-PIECE FORMED WITH

CYLINDERS OF

UNEQUAL DIAMETERS.

PARALLEL LINE DEVELOPMENT

Solutions on p.168
All dimensions in millimetres.

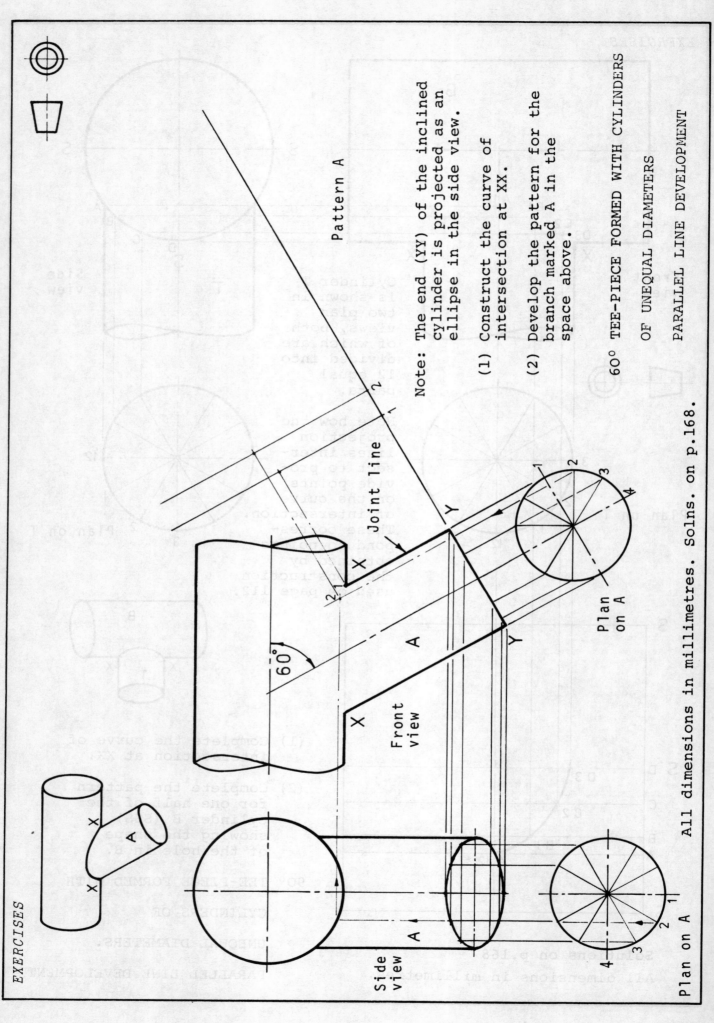

EXERCISES

Pattern A

Note: The end (YY) of the inclined cylinder is projected as an ellipse in the side view.

(1) Construct the curve of intersection at XX.

(2) Develop the pattern for the branch marked A in the space above.

60° TEE-PIECE FORMED WITH CYLINDERS

OF UNEQUAL DIAMETERS

PARALLEL LINE DEVELOPMENT

Joint line

60°

A

Front view

Side view

Plan on A

Plan on A

All dimensions in millimetres. Solns. on p.168.

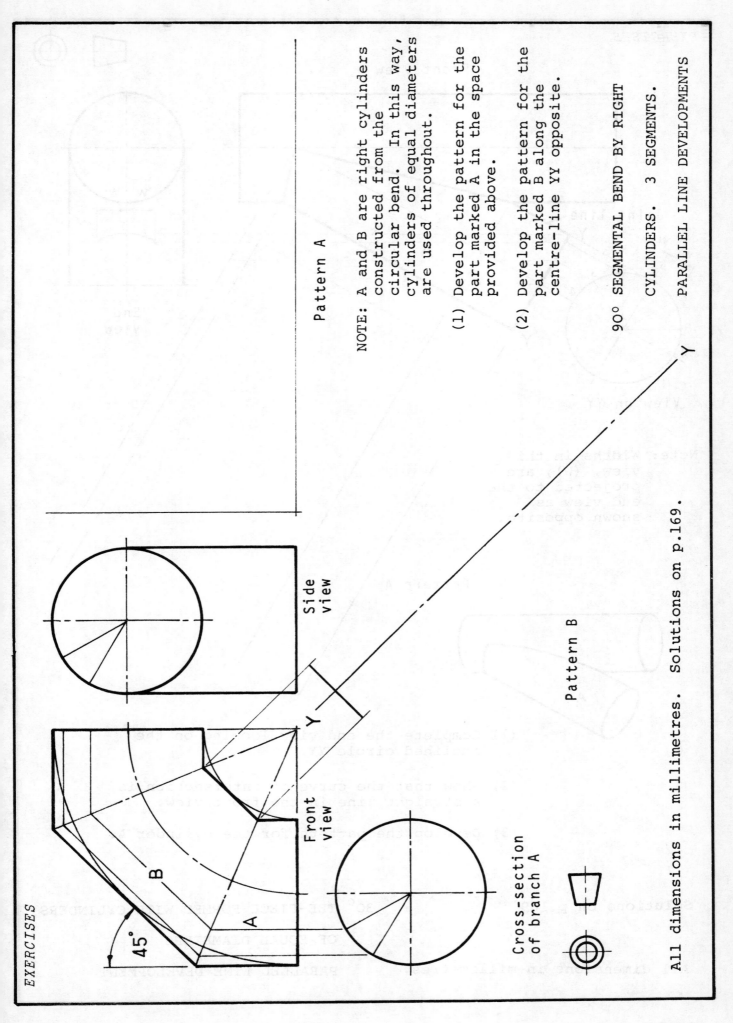

EXERCISES

45°

B

A

Front view

Side view

Y

Cross-section
of branch A

Pattern A

NOTE: A and B are right cylinders
constructed from the
circular bend. In this way,
cylinders of equal diameters
are used throughout.

(1) Develop the pattern for the
part marked A in the space
provided above.

(2) Develop the pattern for the
part marked B along the
centre-line YY opposite.

90° SEGMENTAL BEND BY RIGHT

CYLINDERS. 3 SEGMENTS.

PARALLEL LINE DEVELOPMENTS

Pattern B

Y

All dimensions in millimetres. Solutions on p.169.

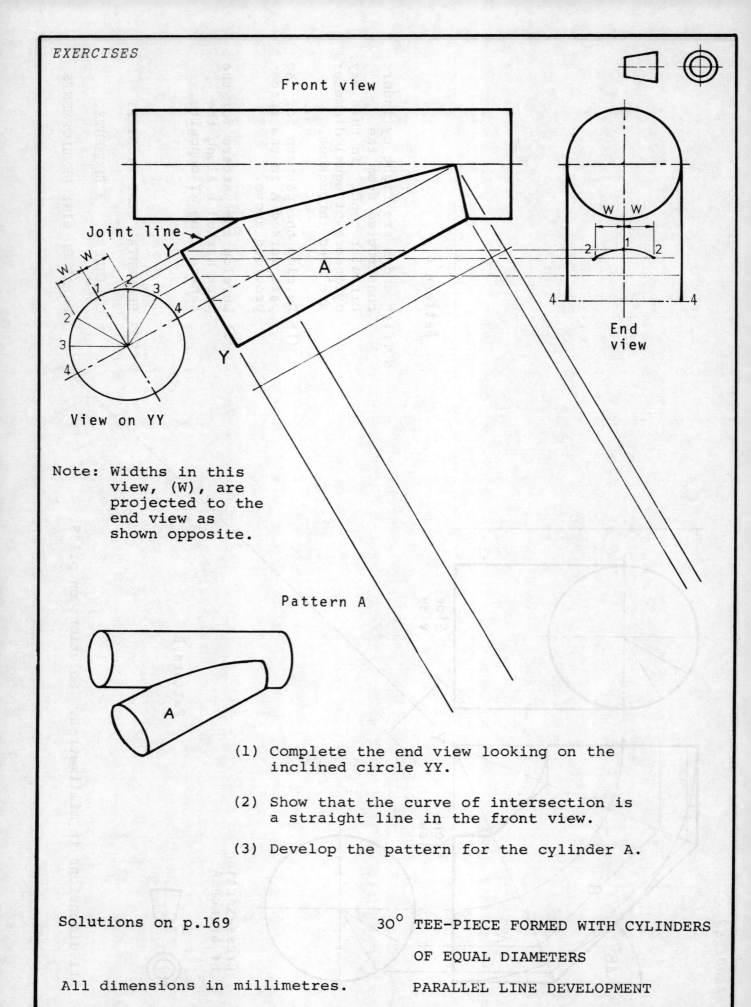

Front view

Joint line

Y

Y

View on YY

A

W W

2
1
3
4

Note: Widths in this
view, (W), are
projected to the
end view as
shown opposite.

W W

2   1   2

4   4

End
view

Pattern A

A

(1) Complete the end view looking on the
    inclined circle YY.

(2) Show that the curve of intersection is
    a straight line in the front view.

(3) Develop the pattern for the cylinder A.

Solutions on p.169

30° TEE-PIECE FORMED WITH CYLINDERS

OF EQUAL DIAMETERS

All dimensions in millimetres.

PARALLEL LINE DEVELOPMENT

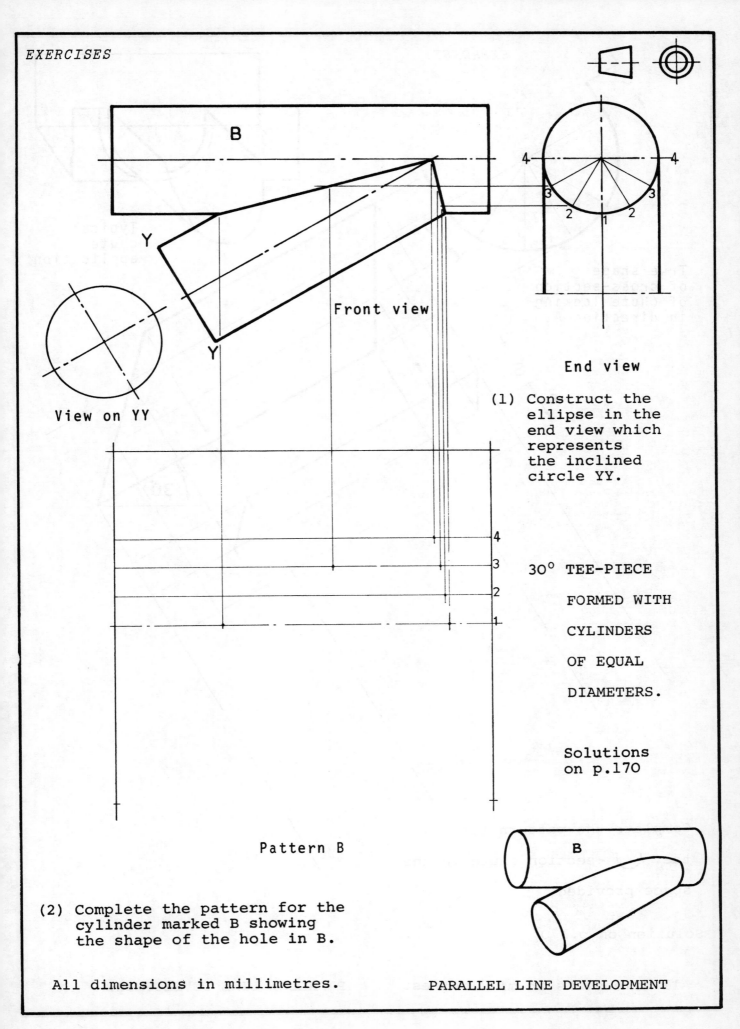

B

Front view

Y

Y

View on YY

End view

4 — 4

3 3

2 1 2

(1) Construct the
ellipse in the
end view which
represents
the inclined
circle YY.

4

3

2

1

30° TEE-PIECE

FORMED WITH

CYLINDERS

OF EQUAL

DIAMETERS.

Solutions
on p.170

Pattern B

B

(2) Complete the pattern for the
cylinder marked B showing
the shape of the hole in B.

All dimensions in millimetres.

PARALLEL LINE DEVELOPMENT

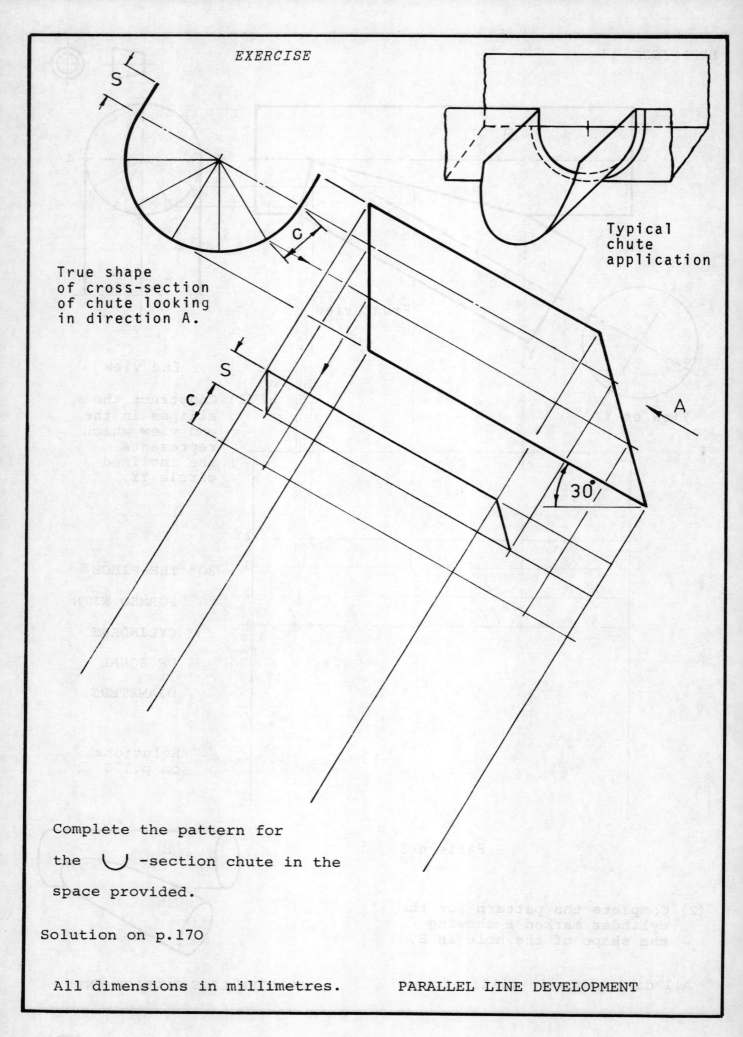

EXERCISE

Typical chute application

True shape
of cross-section
of chute looking
in direction A.

S

C

S

C

30°

A

Complete the pattern for

the ∪ -section chute in the

space provided.

Solution on p.170

All dimensions in millimetres.

PARALLEL LINE DEVELOPMENT

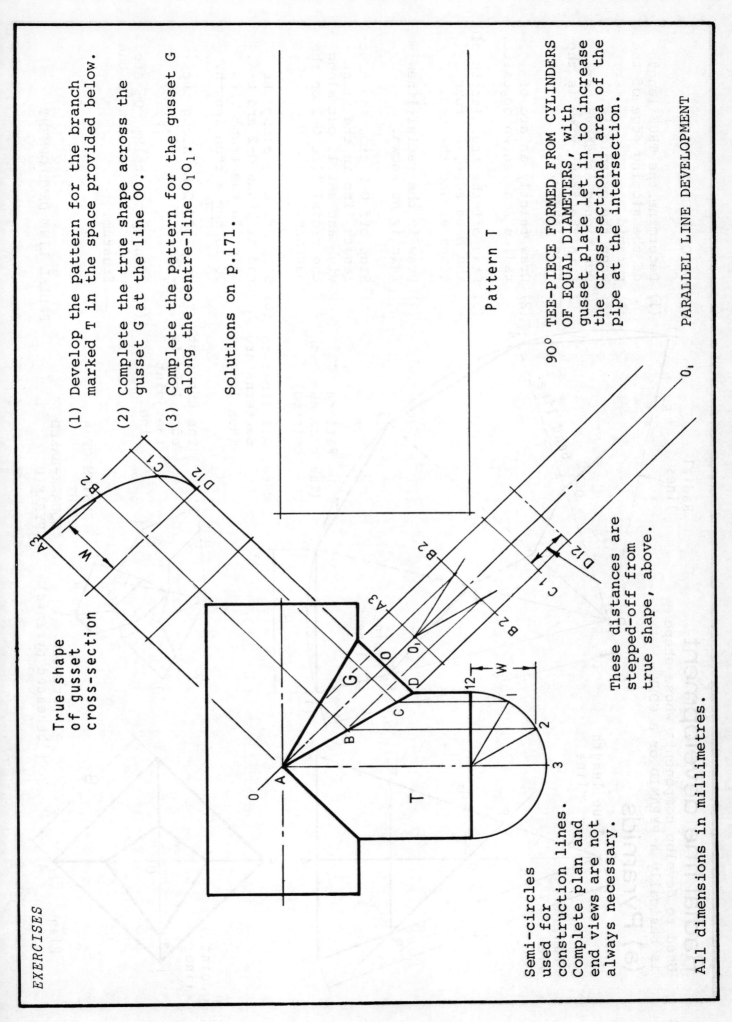

EXERCISES

(1) Develop the pattern for the branch marked T in the space provided below.

(2) Complete the true shape across the gusset G at the line OO.

(3) Complete the pattern for the gusset G along the centre-line O_1O_1.

Solutions on p.171.

Pattern T

90° TEE-PIECE FORMED FROM CYLINDERS OF EQUAL DIAMETERS, with gusset plate let in to increase the cross-sectional area of the pipe at the intersection.

PARALLEL LINE DEVELOPMENT

True shape of gusset cross-section

These distances are stepped-off from true shape, above.

Semi-circles used for construction lines. Complete plan and end views are not always necessary.

All dimensions in millimetres.

119

Radial line development

Used to develop components whose shape is basically a PYRAMID or a CONE

(a) Pyramids

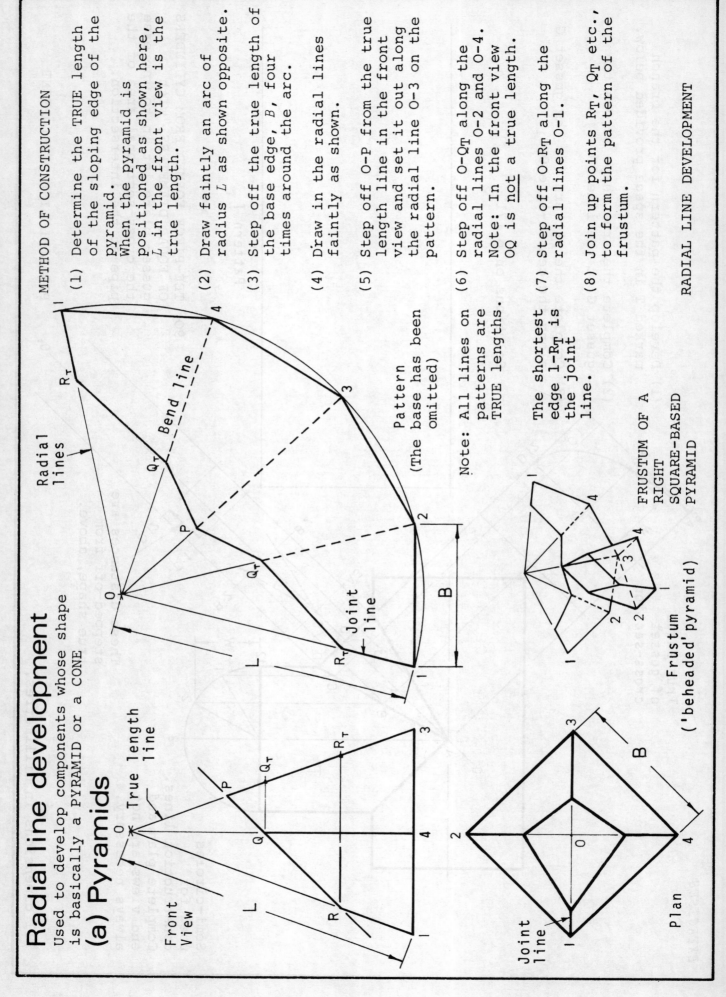

METHOD OF CONSTRUCTION

(1) Determine the TRUE length of the sloping edge of the pyramid.
When the pyramid is positioned as shown here, L in the front view is the true length.

(2) Draw faintly an arc of radius L as shown opposite.

(3) Step off the true length of the base edge, B, four times around the arc.

(4) Draw in the radial lines faintly as shown.

(5) Step off O–P from the true length line in the front view and set it out along the radial line O–3 on the pattern.

(6) Step off O–Q_T along the radial lines O–2 and O–4.
Note: In the front view OQ is not a true length.

(7) Step off O–R_T along the radial lines O–1.

(8) Join up points R_T, Q_T etc., to form the pattern of the frustum.

Note: All lines on patterns are TRUE lengths.

The shortest edge 1–R_T is the joint line.

Radial lines

Bend line

Joint line

Pattern
(The base has been omitted)

FRUSTUM OF A RIGHT SQUARE–BASED PYRAMID

Frustum ('beheaded' pyramid)

RADIAL LINE DEVELOPMENT

Front View

True length line

Joint line

Plan

120

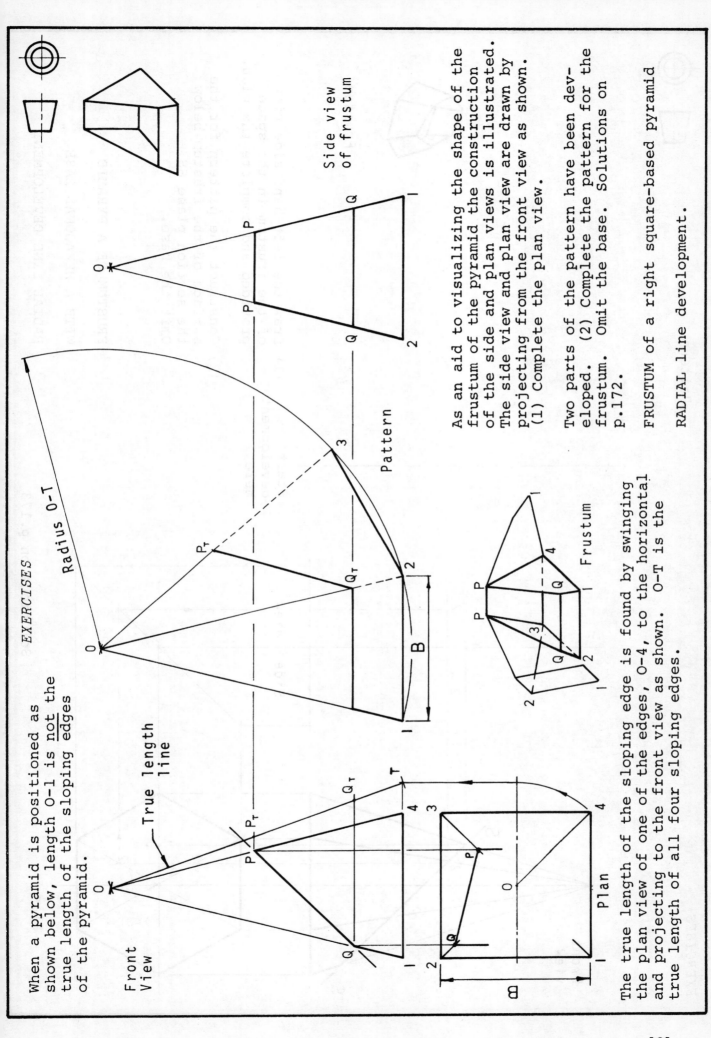

When a pyramid is positioned as shown below, length O-1 is not the true length of the sloping edges of the pyramid.

Front View

True length line

Radius O-T

Plan

B

The true length of the sloping edge is found by swinging the plan view of one of the edges, O-4, to the horizontal and projecting to the front view as shown. O-T is the true length of all four sloping edges.

Side view of frustum

Pattern

Frustum

As an aid to visualizing the shape of the frustum of the pyramid the construction of the side and plan views is illustrated. The side view and plan view are drawn by projecting from the front view as shown.
(1) Complete the plan view.

Two parts of the pattern have been developed. (2) Complete the pattern for the frustum. Omit the base. Solutions on p.172.

FRUSTUM of a right square-based pyramid

RADIAL line development.

121

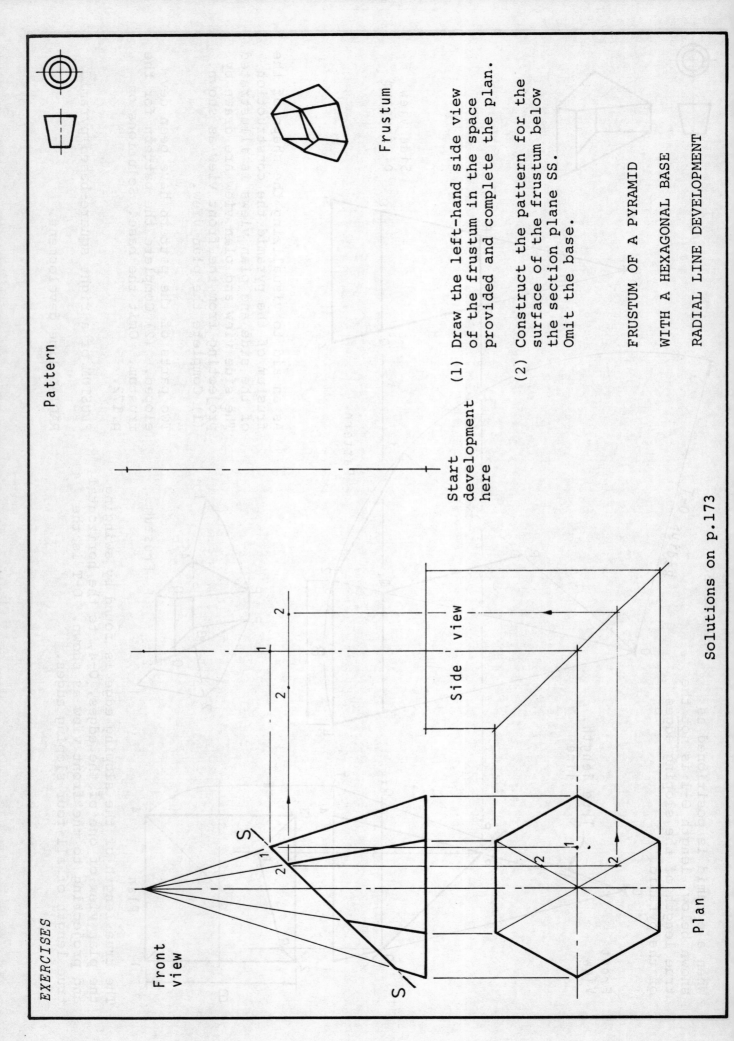

EXERCISES

Pattern

Frustum

Front view

Side view

S

S

Plan

Start development here

(1) Draw the left-hand side view of the frustum in the space provided and complete the plan.

(2) Construct the pattern for the surface of the frustum below the section plane SS. Omit the base.

FRUSTUM OF A PYRAMID

WITH A HEXAGONAL BASE

RADIAL LINE DEVELOPMENT

Solutions on p.173

122

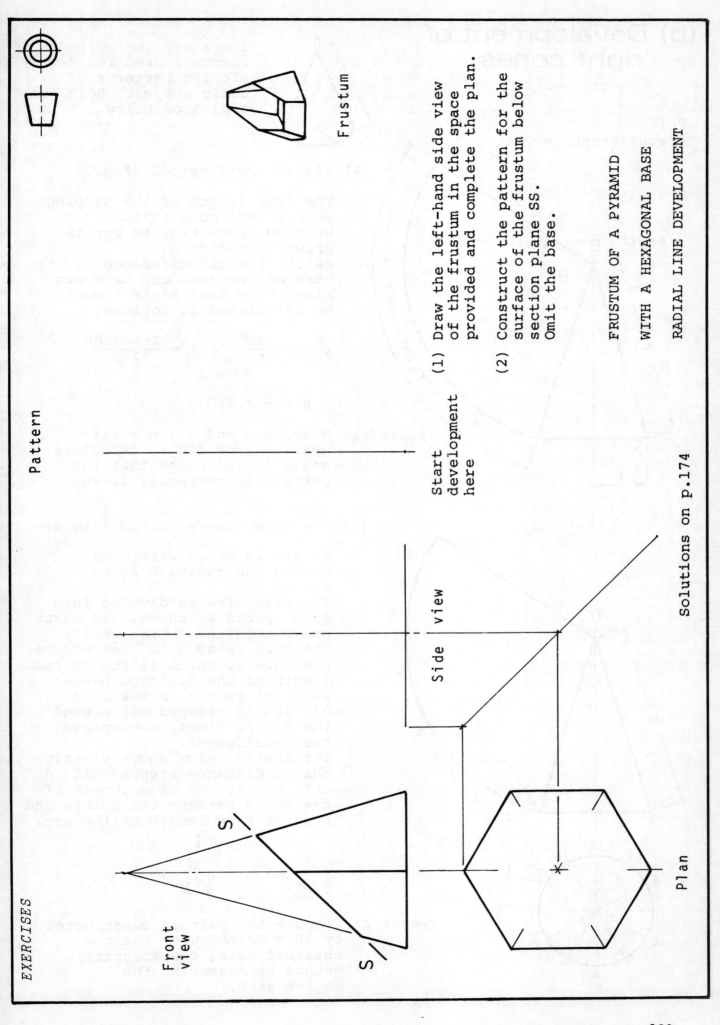

EXERCISES

Pattern

Frustum

(1) Draw the left-hand side view of the frustum in the space provided and complete the plan.

(2) Construct the pattern for the surface of the frustum below the section plane SS. Omit the base.

FRUSTUM OF A PYRAMID

WITH A HEXAGONAL BASE

RADIAL LINE DEVELOPMENT

Start development here

Side view

Solutions on p.174

Front view

Plan

123

(b) Development of right cones

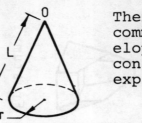

There are two methods commonly used for developing patterns of conic shapes. Both are explained below.

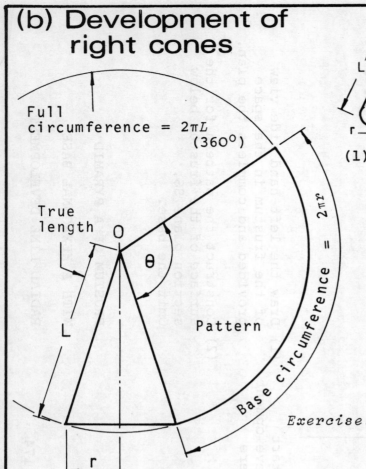

Full circumference = $2\pi L$ (360°)

True length

O

θ

Pattern

L

r

Base circumference = $2\pi r$

Fig.1

(1) *The accurate method* (Fig.1)

The true length of the sloping edge of the cone is L.
Using O as centre, an arc is drawn of radius L.
Before the circumference of the base of the cone can be drawn along this arc, angle θ must be calculated as follows.

$$\frac{\theta}{2\pi r} = \frac{360°}{2\pi L} \qquad \theta = \frac{2\pi r}{2\pi L} 360°$$

$$\theta = \frac{r}{L} \times 360°$$

Exercise: Measure r and L (in millimetres) from fig.1, calculate angle θ, and check that the pattern is correctly drawn.

(2) *The approximate method* (Fig.2)

An arc is drawn with O as centre and radius L as in method 1.
The plan view is divided into equal parts as shown. 12 parts are convenient, being easily obtained using a 30° set-square.
Distance d, which is the approx. length of the distance between pairs of points on the base circle, is stepped off around the arc 12 times to complete the development.
The sketch below shows clearly that a distance stepped off in this way is the true length of the chord between the points and not the true length of the arc.

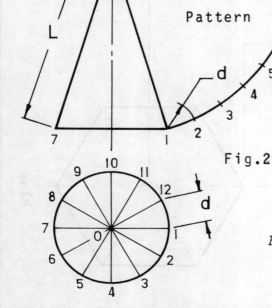

Pattern

L

O

d

Fig.2

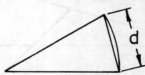

Exercise: Compare the pattern constructed by this method with the one obtained using the accurate method by measuring the angles at O.

DEVELOPMENT OF THE FRUSTUM OF A RIGHT CONE

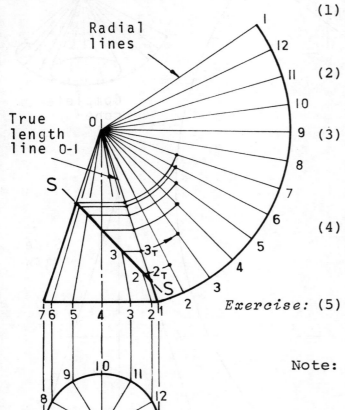

(1) Develop the full surface area by using either the approximate method or the accurate method.

(2) Add radial lines as shown. These correspond to the radial lines drawn on the surface of the cone.

(3) From the points where the radial lines on the surface of the cone intersect the cutting plane SS, project horizontally to the true length line e.g. 2 to 2_T, 3 to 3_T etc.

(4) Swing these true lengths from O to the corresponding radial lines on the pattern, e.g. 2_T to line 2, 3_T to line 3 etc.

Exercise: (5) Complete the pattern by adding the remaining points and joining with a heavy line.

Note: All lengths on the pattern must be true lengths.
The shortest edge is usually used as the joint line. In this case it is along line O-1.

As an aid to visualizing the shape of conic frusta the construction of the front and plan views is illustrated and explained below.

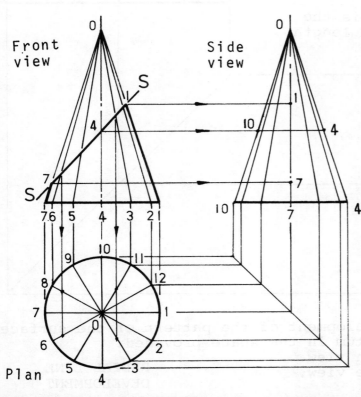

Construction of plan view and side view of a conic frustum

Plan view
Project the points of intersection of the radial lines and the cutting plane S-S from the front view to the corresponding radial lines in the plan as shown.

e.g. points $\frac{6}{8}$ to lines 6 and 8.

Side view
Project the points of intersection of the radial lines and the cutting plane S-S from the front view to the corresponding radial lines in the side view as shown.

e.g. points $\frac{4}{10}$ to lines 4 and 10.

Exercises: Complete the plan and side view.

RADIAL LINE DEVELOPMENT

Solns. p.175

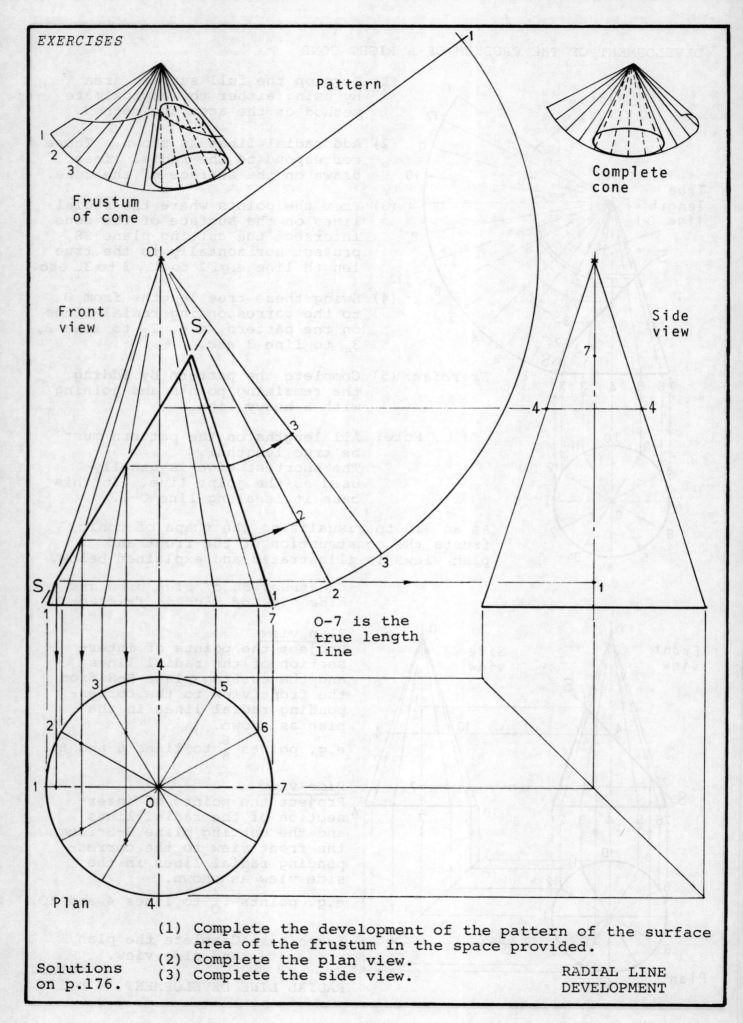

Pattern

Frustum
of cone

Complete
cone

Front
view

S

Side
view

S

O-7 is the
true length
line

Plan

(1) Complete the development of the pattern of the surface
    area of the frustum in the space provided.
(2) Complete the plan view.
(3) Complete the side view.

Solutions
on p.176.

RADIAL LINE
DEVELOPMENT

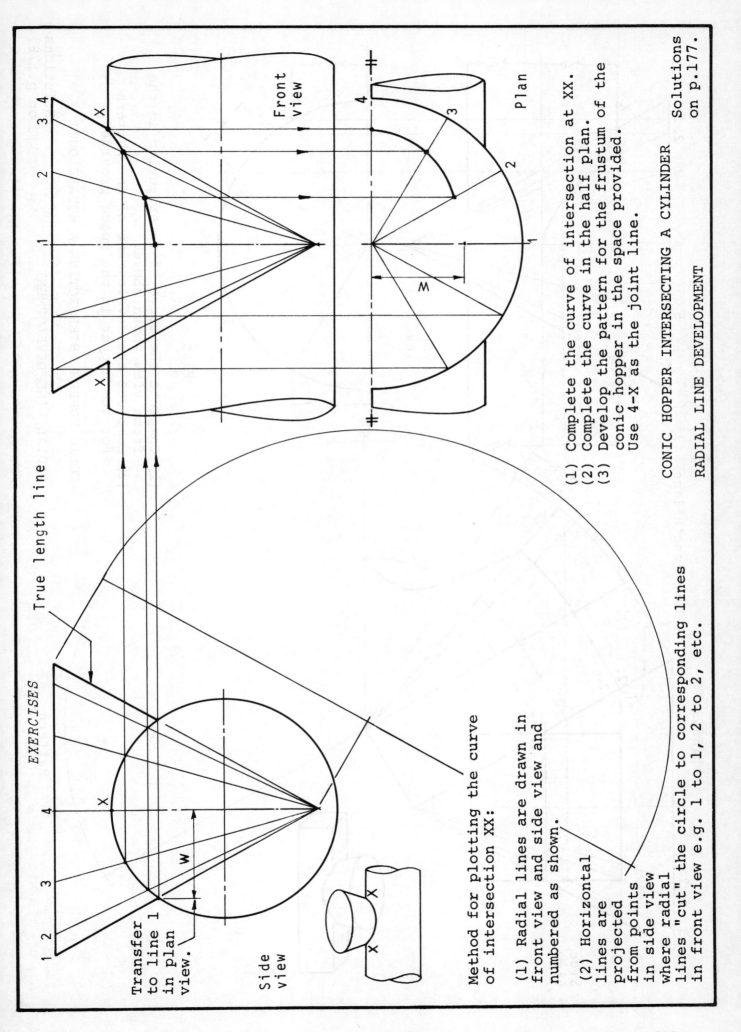

EXERCISES

True length line

Transfer to line 1 in plan view.

Side view

Method for plotting the curve of intersection XX:

(1) Radial lines are drawn in front view and side view and numbered as shown.

(2) Horizontal lines are projected from points in side view where radial lines "cut" the circle to corresponding lines in front view e.g. 1 to 1, 2 to 2, etc.

Front view

Plan

(1) Complete the curve of intersection at XX.
(2) Complete the curve in the half plan.
(3) Develop the pattern for the frustum of the conic hopper in the space provided. Use 4-X as the joint line.

CONIC HOPPER INTERSECTING A CYLINDER

RADIAL LINE DEVELOPMENT

Solutions on p.177.

127

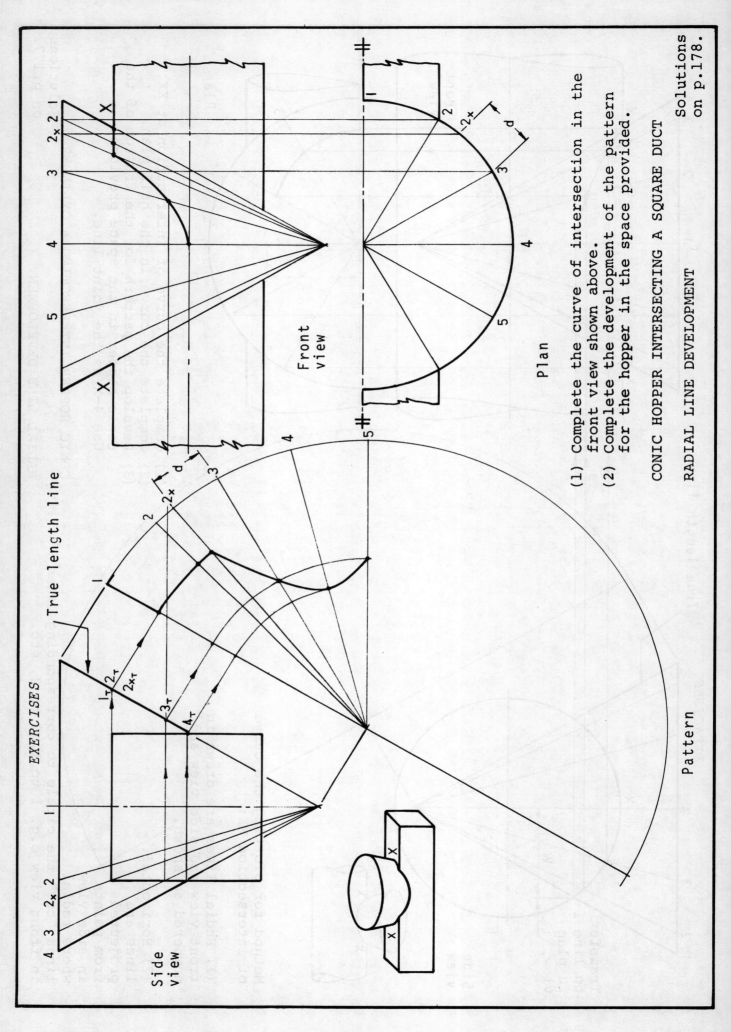

EXERCISES

Side view

True length line

Front view

Plan

Pattern

(1) Complete the curve of intersection in the front view shown above.

(2) Complete the development of the pattern for the hopper in the space provided.

CONIC HOPPER INTERSECTING A SQUARE DUCT

RADIAL LINE DEVELOPMENT

Solutions on p.178.

128

THE COMMON CENTRAL SPHERE – A SPECIAL CASE

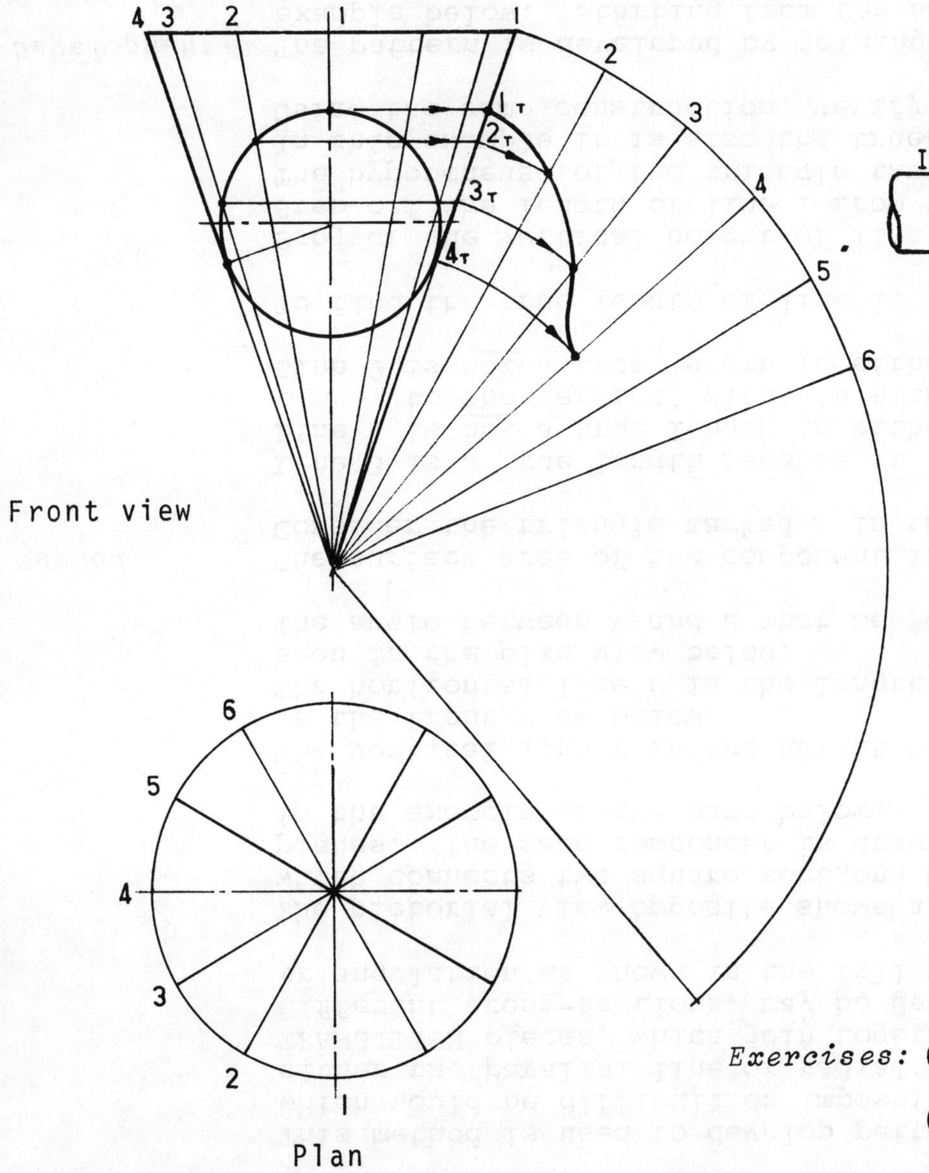

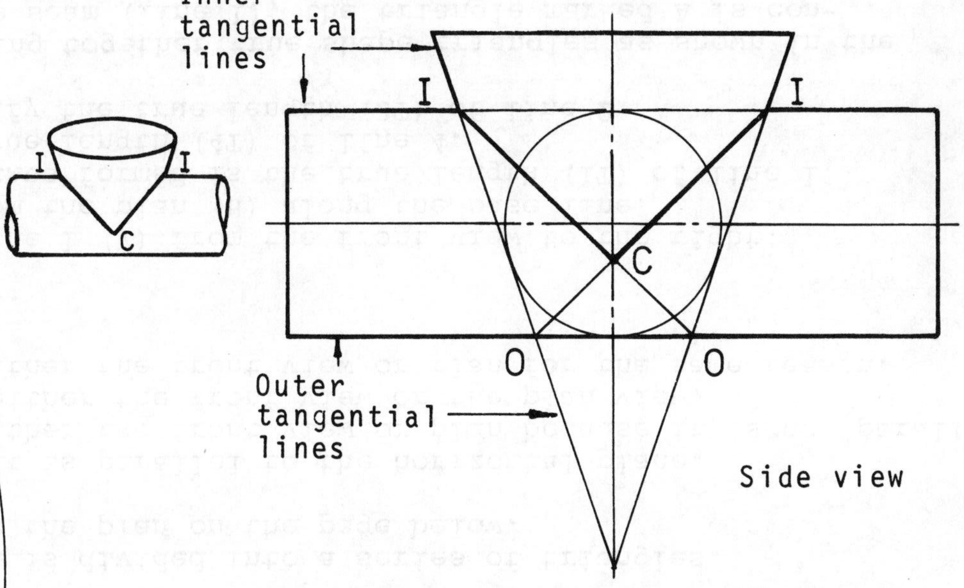

Front view

Side view

Inner tangential lines

Outer tangential lines

When a circle (an imaginary sphere) can be drawn within two intersecting components so that both outside edges are tangential to the circle as shown above, a special construction may be used to determine the "curve of intersection".
Intersection points of the inner tangential lines (I) are joined to the intersection points of the outer tangential lines (O) as shown in the side view above.
The "curve" of intersection I-C-I appears as two straight lines.
Note: C is below the centre-line of the cylinder.

Exercises: (1) Verify this "curve" of intersection by using the radial line method shown previously.
(2) Complete the development of the pattern for the frustum of the cone in the space provided.

Solutions on p.179

RADIAL LINE DEVELOPMENT

Plan

CONICAL HOPPER INTERSECTING A CYLINDER

Triangulation

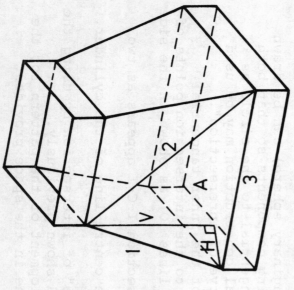

This method is used to develop patterns of components which would be difficult or impossible to develop using either the parallel line or radial line methods. TRANSITION pieces, which join together ducts which have different cross-sections, may be developed using triangulation as shown in the following examples.

The pictorial view opposite shows a transition piece which connects two square sections between parallel planes. The same component is drawn orthographically in the example on the page below.

The vertical line V is the height of line 1 as seen in the front view below.
The horizontal line H is the length of line 1 as seen in the plan view below.
The angle between V and H must be 90°.

Method

The surface area of the component is divided into a series of triangles. Consider the triangle marked A in the plan on the page below.

Line 3 is a true length because it is parallel to the horizontal plane.
Line 1 is not a true length in either the front view or plan because it is not parallel to the vertical plane in either the front view or the plan view.
Line 2 is not a true length in either the front view or plan for the same reason.

To find the true length of line 1.

Project the vertical height of line 1 (V) from the front view to the right.
Step off the length of line 1 from the plan (H) along the base line.
The hypoteneuse of the triangle thus formed is the true length (1T) of line 1.
In this example it is also the true length (4T) of line 4.
Using the same construction, verify the true length (2T) of line 2.

Development

The pattern is developed by joining together true shape triangles as shown in the example below. Starting from the seam (line 1), the triangle marked A is constructed using the true length for each edge. Triangles B, C, D, etc., are then added in the correct sequence as shown.

130

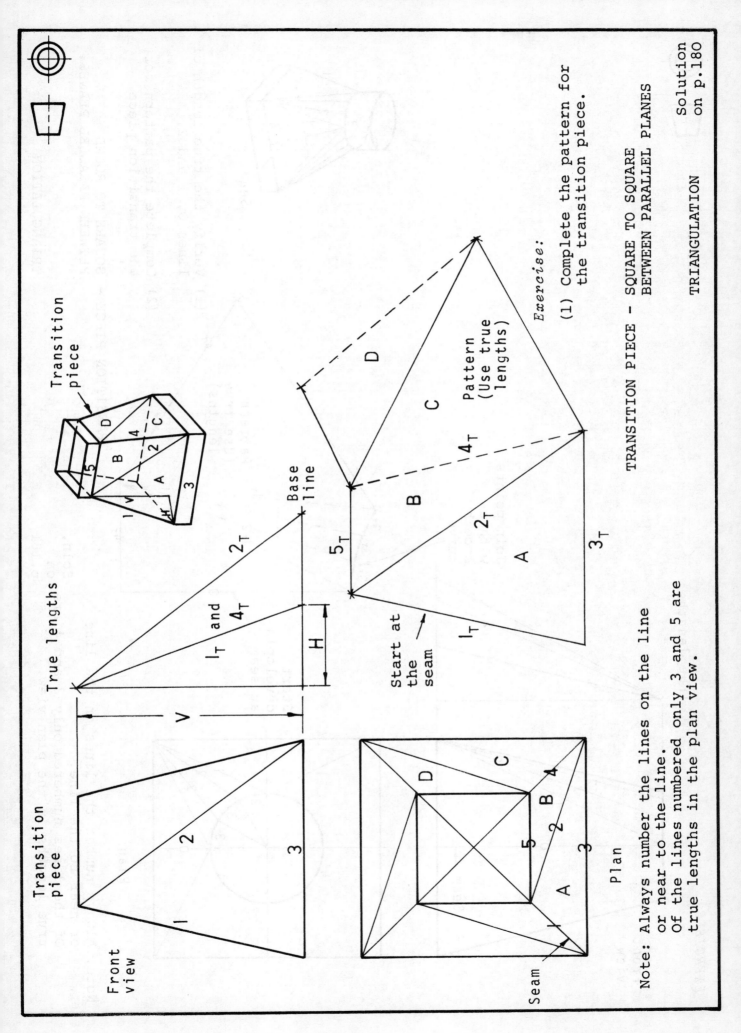

Transition
piece

Transition piece

True lengths

Front view

Base line

Plan

Start at the seam

Seam

Note: Always number the lines on the line or near to the line. Of the lines numbered only 3 and 5 are true lengths in the plan view.

Pattern (Use true lengths)

Exercise:

(1) Complete the pattern for the transition piece.

TRANSITION PIECE – SQUARE TO SQUARE BETWEEN PARALLEL PLANES

Solution on p.180

TRIANGULATION

131

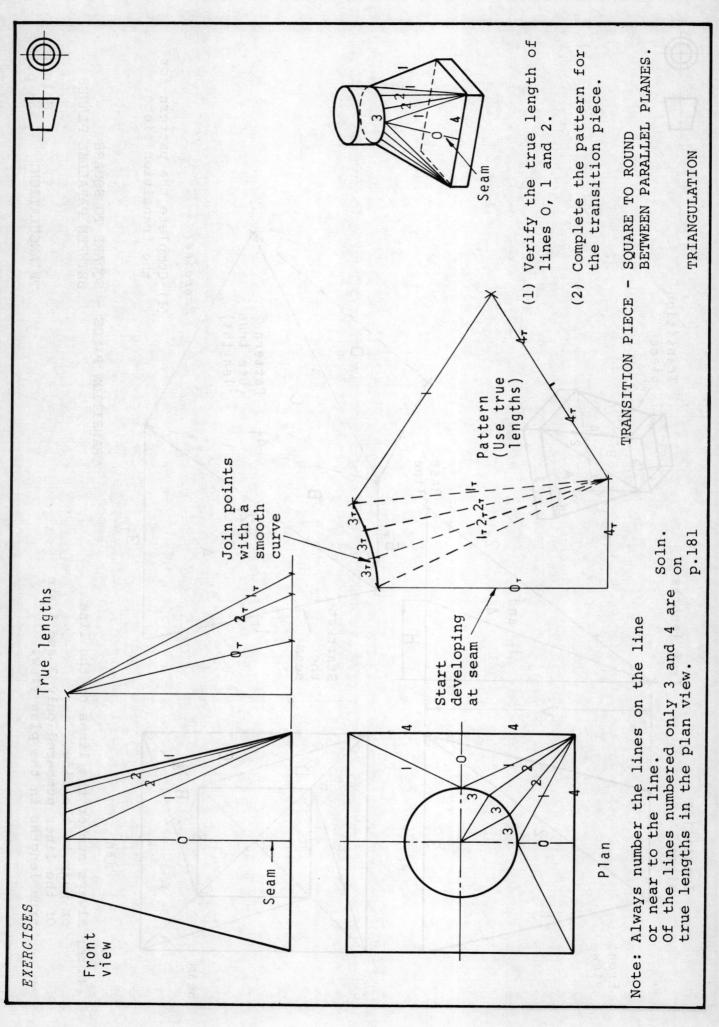

EXERCISES

Front view

True lengths

Join points with a smooth curve

Pattern (Use true lengths)

Start developing at seam

Seam

Plan

Seam

(1) Verify the true length of lines 0, 1 and 2.

(2) Complete the pattern for the transition piece.

TRANSITION PIECE – SQUARE TO ROUND BETWEEN PARALLEL PLANES.

Note: Always number the lines on the line or near to the line. Of the lines numbered only 3 and 4 are true lengths in the plan view.

Soln. on p.181

TRIANGULATION

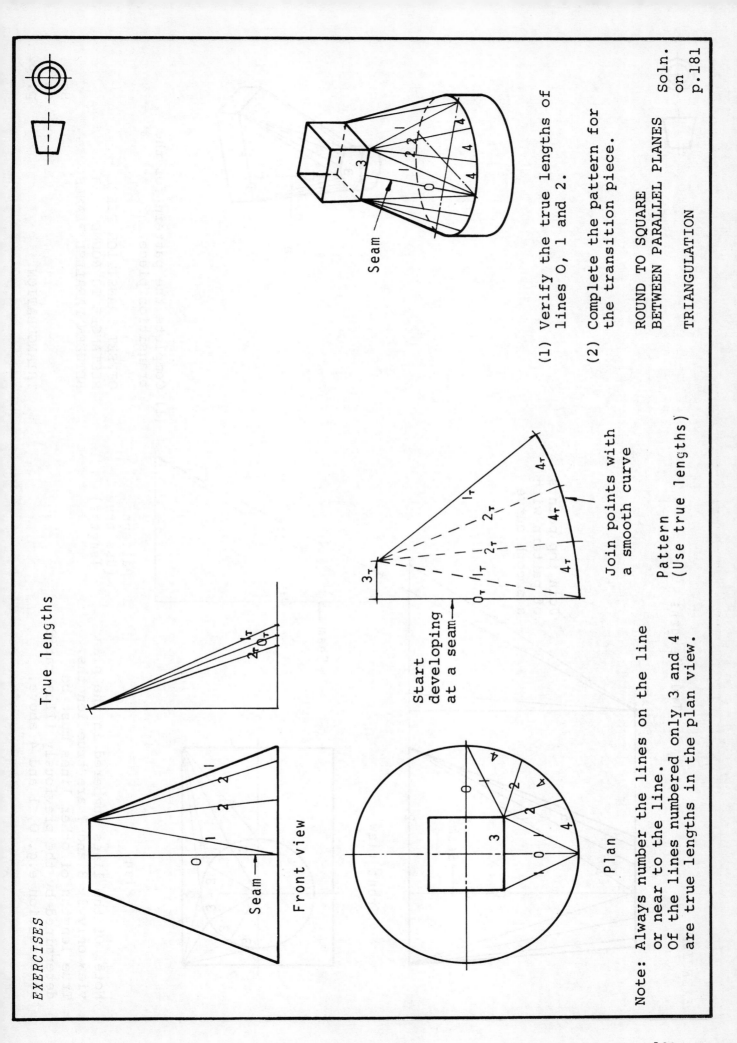

True lengths

Front view

Seam

Plan

Start developing at a seam

Join points with a smooth curve

Pattern
(Use true lengths)

Seam

(1) Verify the true lengths of lines 0, 1 and 2.

(2) Complete the pattern for the transition piece.

ROUND TO SQUARE
BETWEEN PARALLEL PLANES

TRIANGULATION

Soln. on p.181

Note: Always number the lines on the line or near to the line.
Of the lines numbered only 3 and 4 are true lengths in the plan view.

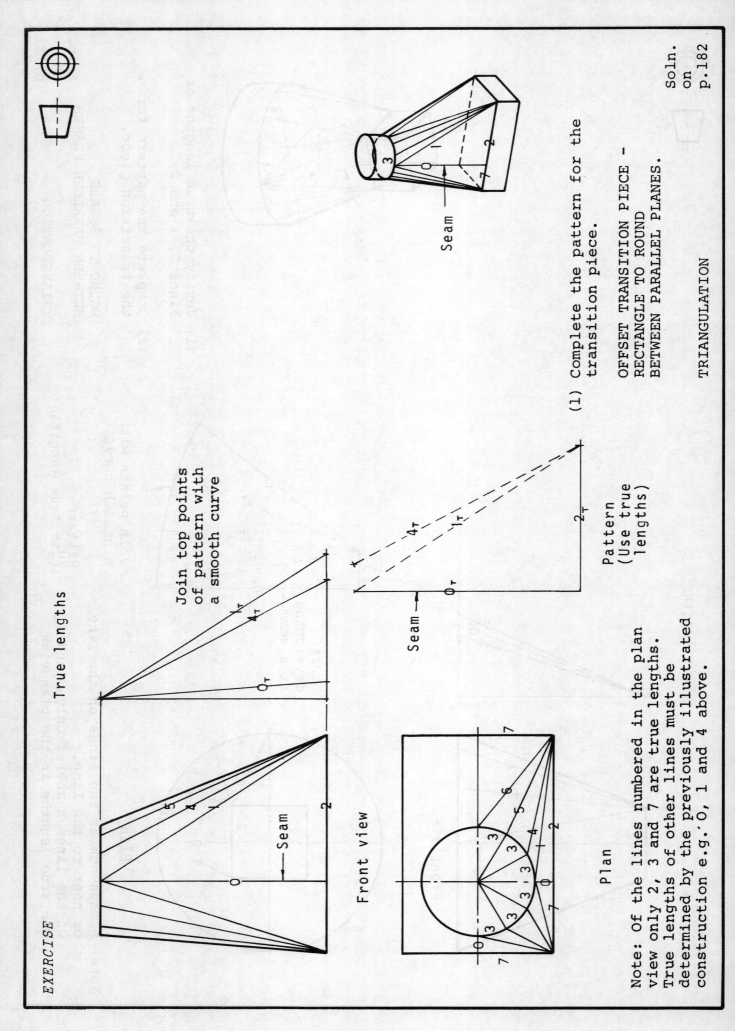

EXERCISE

True lengths

Join top points
of pattern with
a smooth curve

Front view

Plan

Note: Of the lines numbered in the plan
view only 2, 3 and 7 are true lengths.
True lengths of other lines must be
determined by the previously illustrated
construction e.g. 0, 1 and 4 above.

Seam

Pattern
(Use true
lengths)

(1) Complete the pattern for the
transition piece.

OFFSET TRANSITION PIECE –
RECTANGLE TO ROUND
BETWEEN PARALLEL PLANES.

Soln.
on
p.182

TRIANGULATION

134

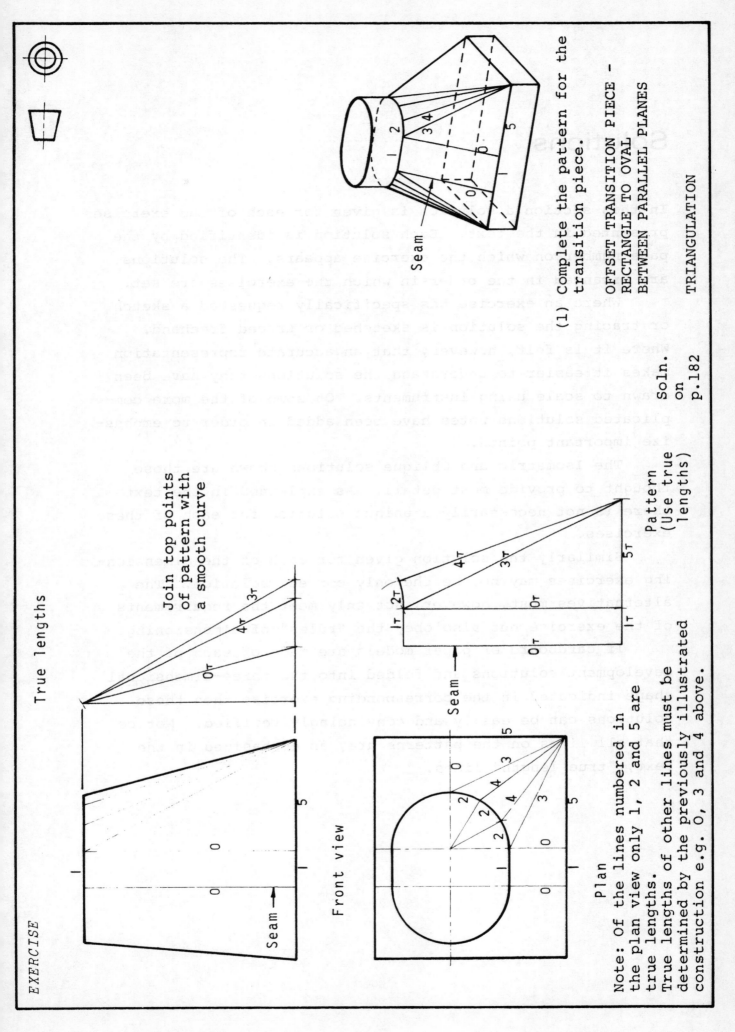

EXERCISE

OFFSET TRANSITION PIECE –
RECTANGLE TO OVAL
BETWEEN PARALLEL PLANES

TRIANGULATION

(1) Complete the pattern for the
transition piece.

Soln.
on
p.182

True lengths

Join top points
of pattern with
a smooth curve

Pattern
(Use true
lengths)

Seam

Front view

Plan

Note: Of the lines numbered in
the plan view only 1, 2 and 5 are
true lengths.
True lengths of other lines must be
determined by the previously illustrated
construction e.g. O, 3 and 4 above.

Seam

135

Solutions

In this section a solution is given for each of the exercises
presented in the text. Each solution is identified by the
page number on which the exercise appears. The solutions
are arranged in the order in which the exercises are set.

Where an exercise has specifically requested a sketch
or tracing the solution is sketched or traced freehand.
Where it is felt, however, that an accurate representation
makes it easier to understand the solutions they have been
drawn to scale using instruments. On some of the more complicated solutions notes have been added in order to emphas-
ize important points.

The Isometric and Oblique solutions shown are those
thought to provide most detail. As explained in the text
there is not necessarily a unique solution for each of these
exercises.

Similarly the solution given for each of the dimensioning exercises may not be the only correct solution. The
alternatives must, however, not only meet the requirements
of the exercise but also obey the "rules" of dimensioning.

If cardboard or paper models are made of each of the
development solutions and folded into the three-dimensional
shape indicated in the corresponding exercise then these
solutions can be easily and convincingly verified. Notice
that all lines on the patterns are, as emphasized in the
text, "true length" lines.

Solutions: First Angle Orthographic Projection

On page 7	Drg.	Reason for incorrect interpretation
	1	Plan incorrectly positioned. It should be below F.
	2	Plan incorrectly positioned. It should be projected below F.
	3	Plan and side view incorrectly positioned. They should be interchanged.
	4	Side view out of position. It should be projected from F.
	5	Plan incorrectly positioned. It should be projected below F.
	6	Plan incorrectly drawn. The cut away should be at the front of P.
	7	Side view incorrectly drawn. The cut away should be hidden detail.
	8	Side view positioned incorrectly. It should be drawn on the right hand side of F.
	9	Plan drawn incorrectly. The cut away should be at the rear of P.

On page 9

1	C
2	D
3	K
4	L
5	F
6	H
7	A
8	B
9	G
10	J
11	I
12	E

On page 8   Note   These are sketched freehand.

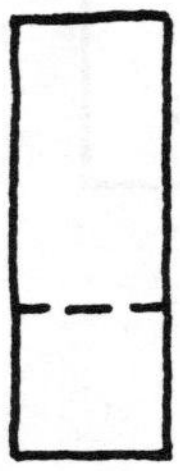

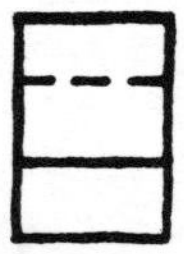

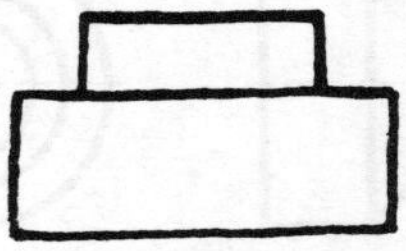

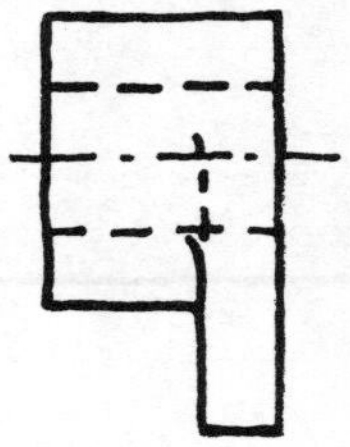

On page 10

Drawing	A	B	C	D	E	F
Front view in direction of F	10	1	11	4	7	6
Plan view in direction of P	14	17	8	3	18	9
Side view in direction of R	5	16	2	12	13	15

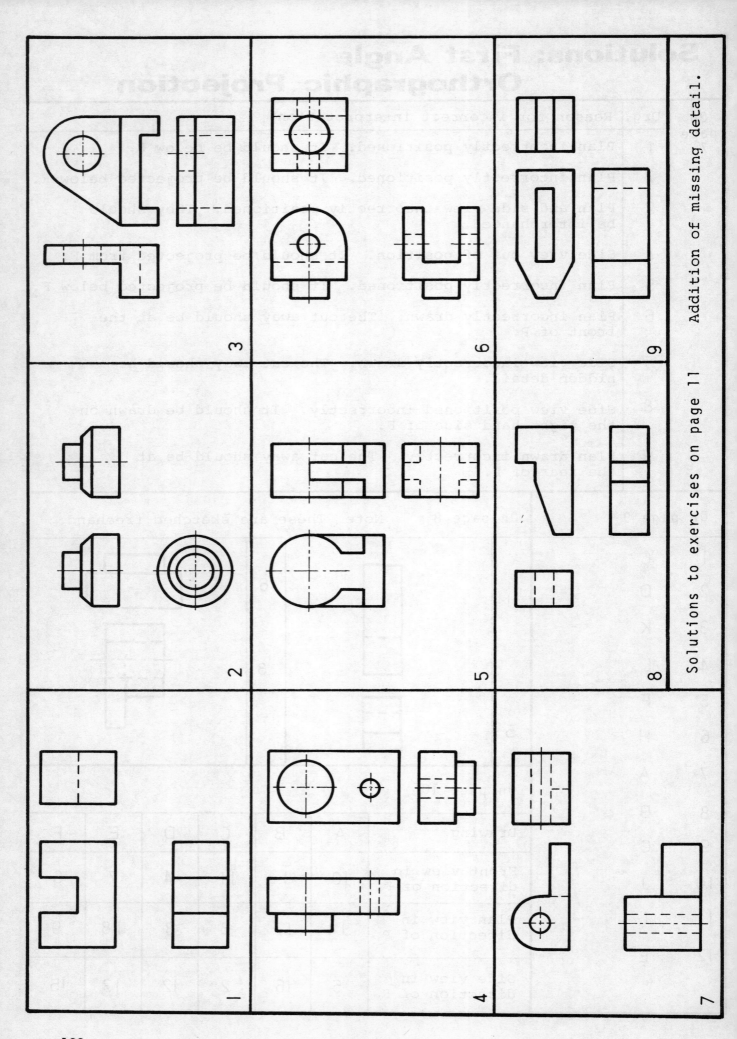

Solutions: First Angle Orthographic Projection

Addition of missing detail.

Solutions to exercises on page 11

3

6

9

2

5

8

1

4

7

138

Solutions to exercises on pages 12 and 13 (sketched freehand)

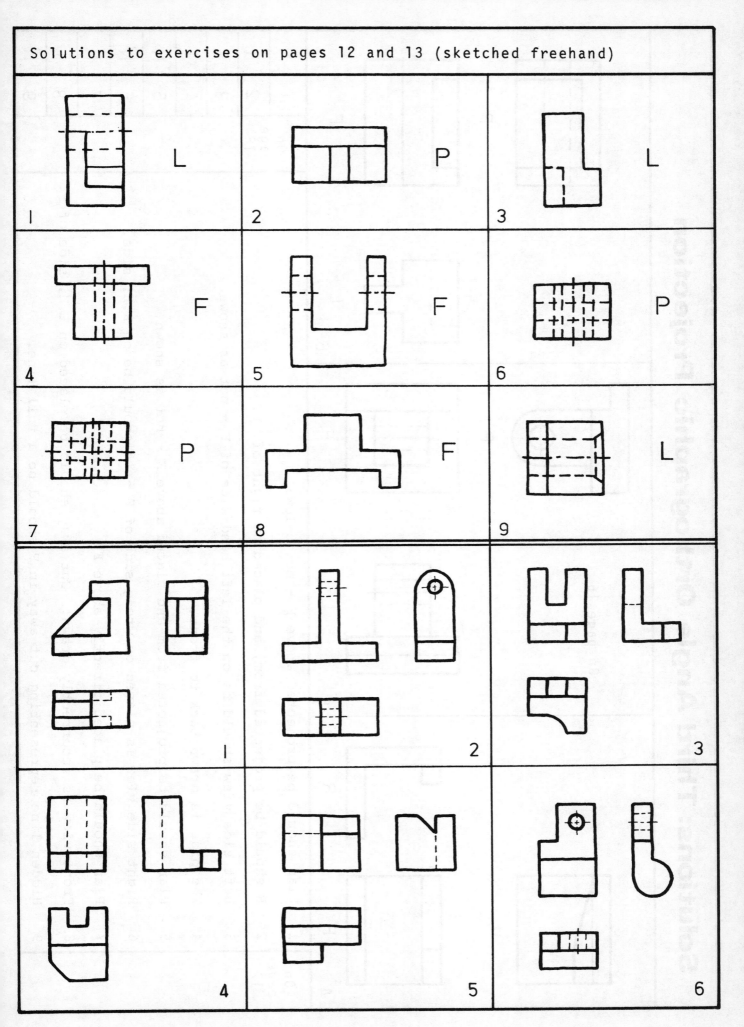

Solutions: Third Angle Orthographic Projection

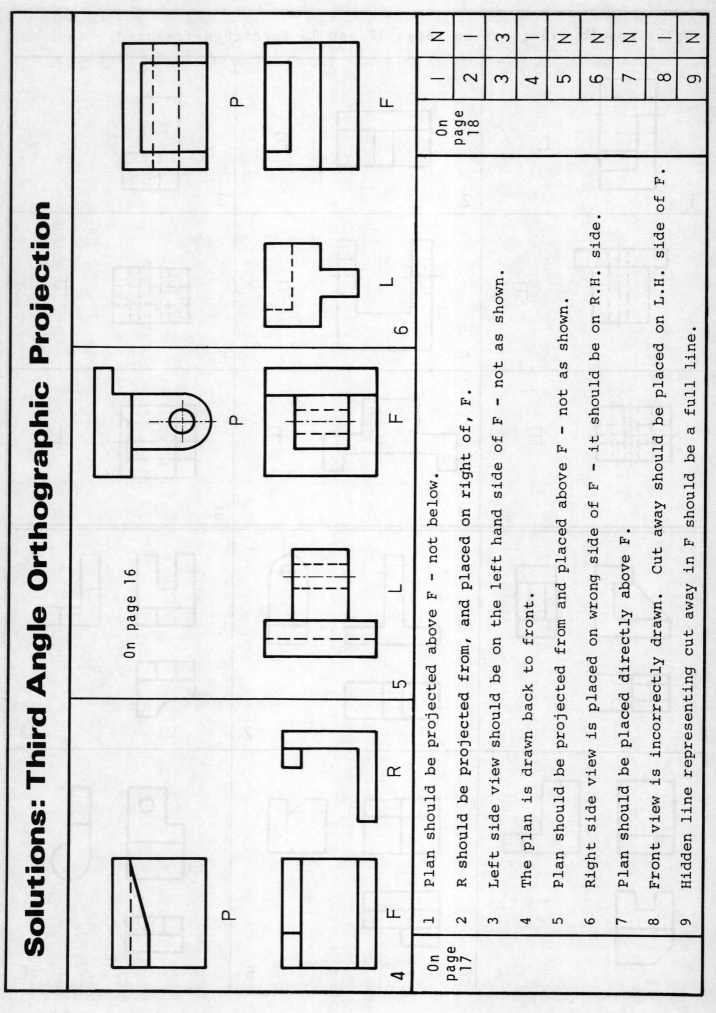

On page 18		
	1	N
	2	I
	3	3
	4	I
	5	N
	6	N
	7	N
	8	I
	9	N

On page 17	
1	Plan should be projected above F – not below.
2	R should be projected from, and placed on right of, F.
3	Left side view should be on the left hand side of F – not as shown.
4	The plan is drawn back to front.
5	Plan should be projected from and placed above F – not as shown.
6	Right side view is placed on wrong side of F – it should be on R.H. side.
7	Plan should be placed directly above F.
8	Front view is incorrectly drawn. Cut away should be placed on L.H. side of F.
9	Hidden line representing cut away in F should be a full line.

On page 16

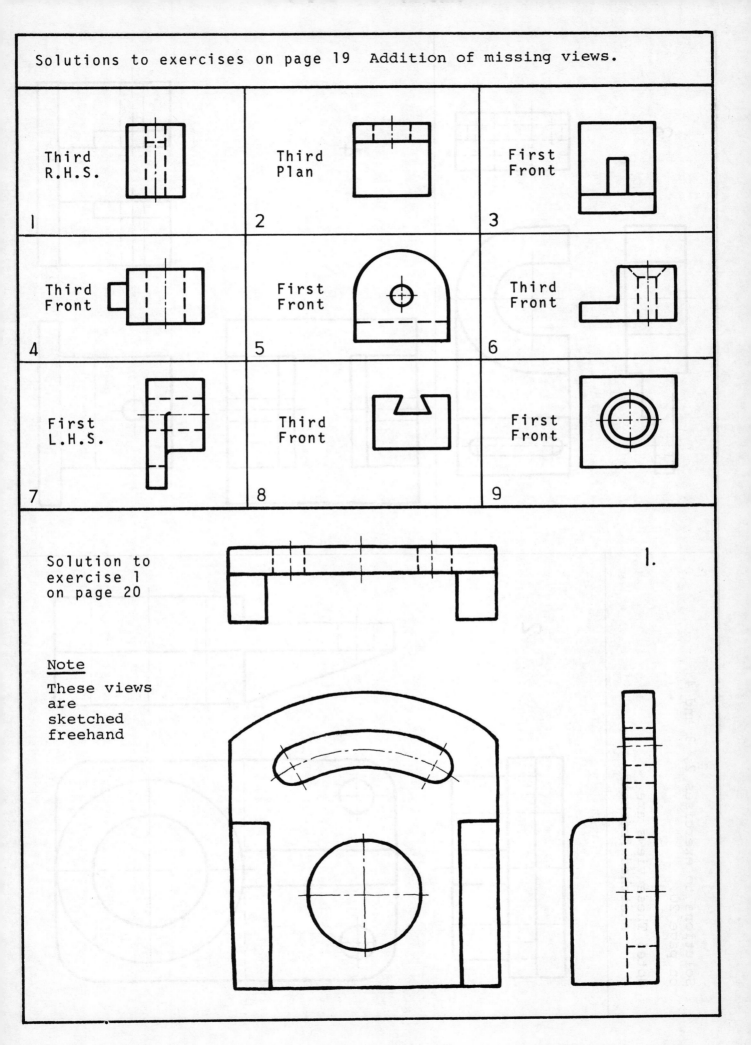

Solutions to exercises on page 19   Addition of missing views.

1 Third R.H.S.

2 Third Plan

3 First Front

4 Third Front

5 First Front

6 Third Front

7 First L.H.S.

8 Third Front

9 First Front

Solution to exercise 1 on page 20

1.

<u>Note</u>

These views are sketched freehand

141

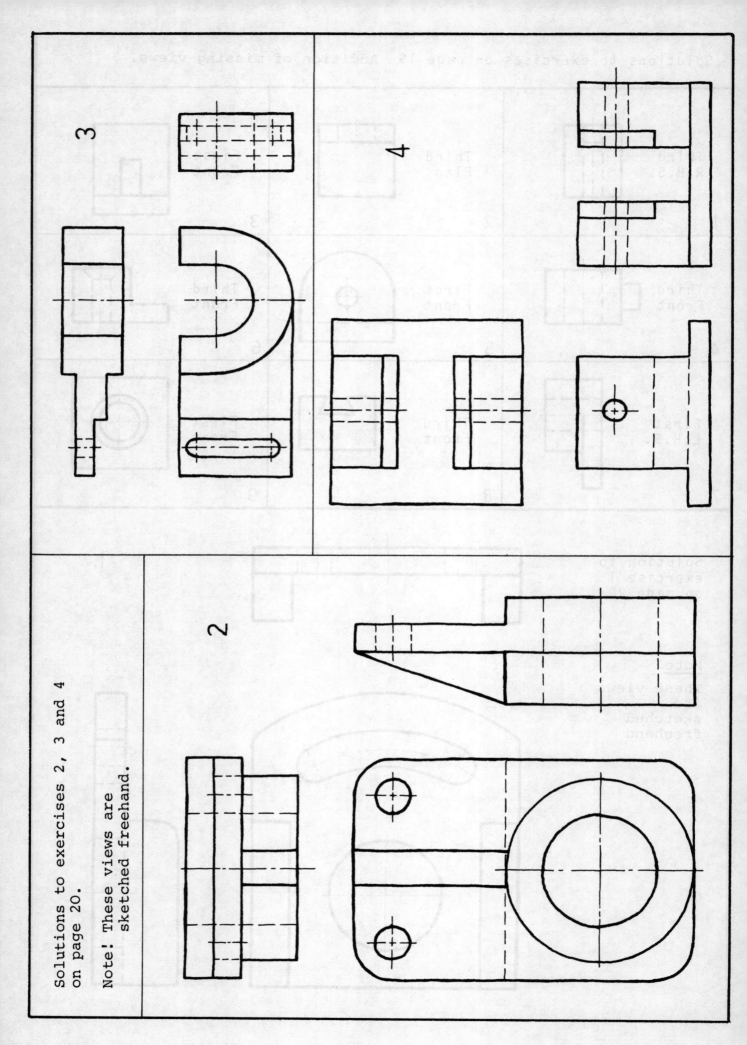

Solutions to exercises 2, 3 and 4 on page 20.

Note! These views are sketched freehand.

2

3

4

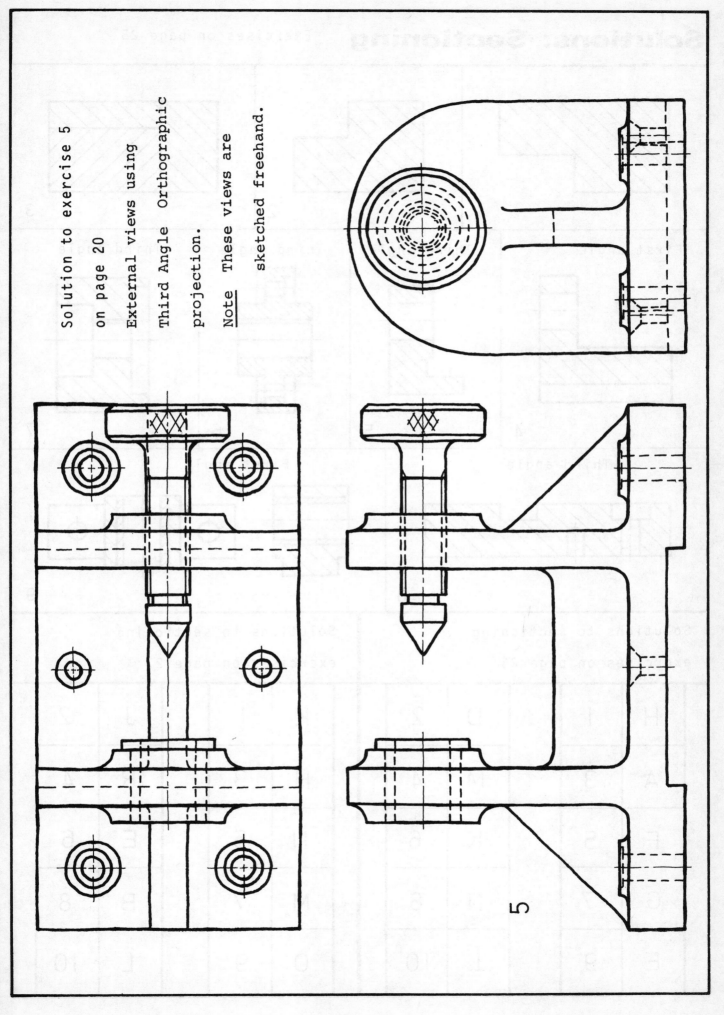

Solution to exercise 5
on page 20

External views using

Third Angle Orthographic

projection.

Note These views are

sketched freehand.

5

Solutions: Sectioning

Exercises on page 25

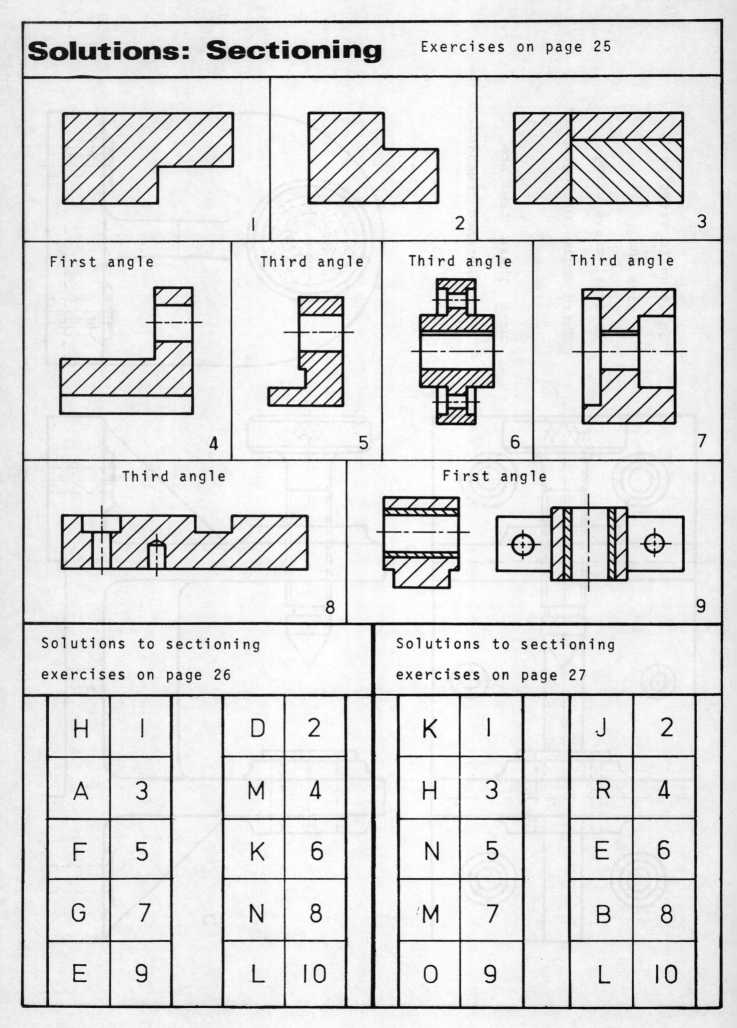

1

2

3

First angle

Third angle

Third angle

Third angle

4

5

6

7

Third angle

First angle

8

9

Solutions to sectioning
exercises on page 26

H	I		D	2
A	3		M	4
F	5		K	6
G	7		N	8
E	9		L	10

Solutions to sectioning
exercises on page 27

K	I		J	2
H	3		R	4
N	5		E	6
M	7		B	8
O	9		L	10

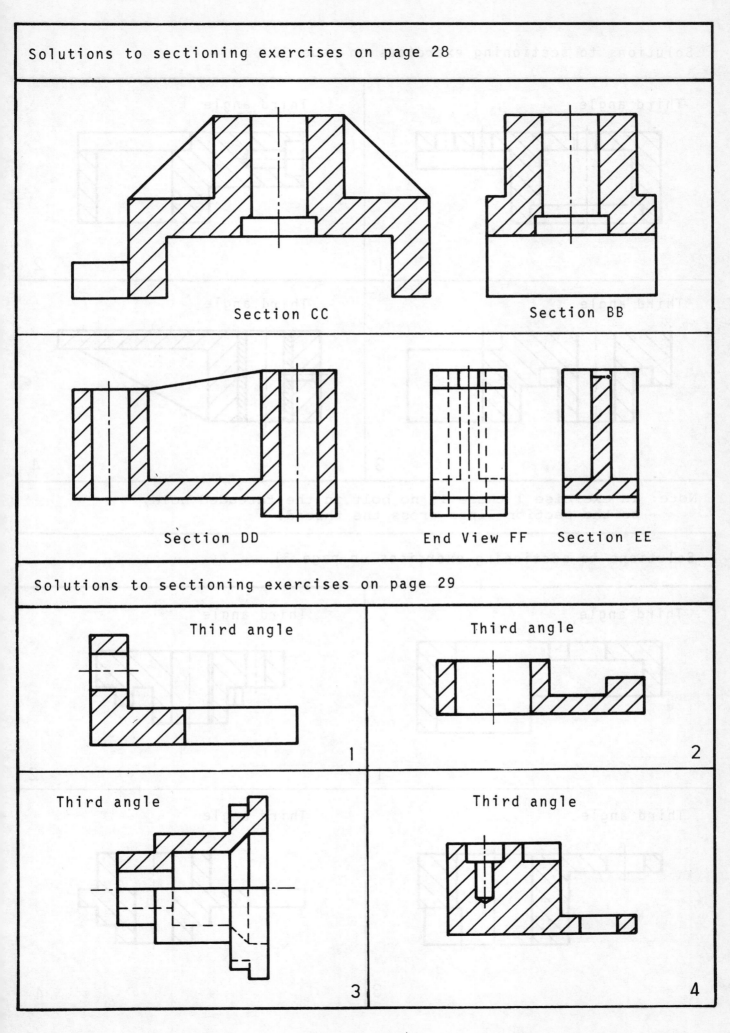

Solutions to sectioning exercises on page 28

Section CC

Section BB

Section DD

End View FF   Section EE

Solutions to sectioning exercises on page 29

Third angle

1

Third angle

2

Third angle

3

Third angle

4

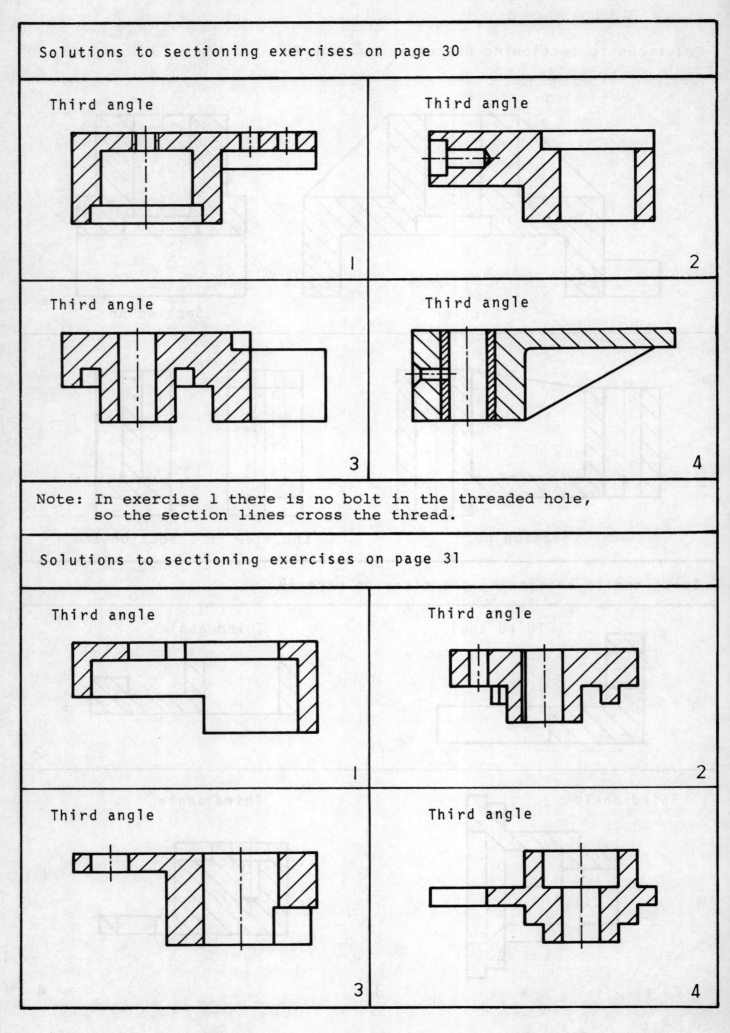

Solutions to sectioning exercises on page 30

Third angle

1

Third angle

2

Third angle

3

Third angle

4

Note: In exercise 1 there is no bolt in the threaded hole,
so the section lines cross the thread.

Solutions to sectioning exercises on page 31

Third angle

1

Third angle

2

Third angle

3

Third angle

4

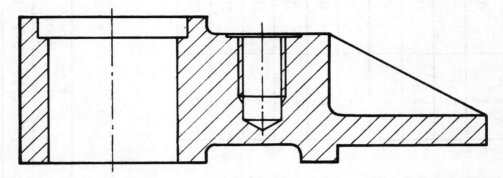

A sectional Front View looking on cutting plane CC

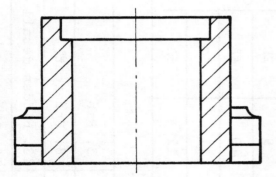

A sectional Side View looking
on cutting plane AA

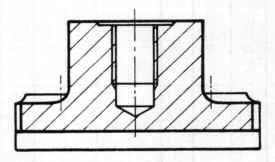

A sectional Side View looking
on cutting plane BB

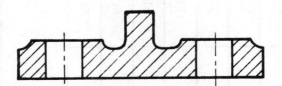

A sectional Side View looking
on cutting plane DD

Third Angle Orthographic projection

Solutions to exercises on page 45

TERMINOLOGY

1	5D
2	4D
3	6B
4	1C
5	1B
6	3A
7	1A
8	5B
9	6C
10	5C

Solutions to exercises on page 46

ABBREVIATIONS

1	4C
2	1A
3	1D
4	6C
5	4B
6	5D
7	1B
8	2C
9	2D
10	4A
11	5C
12	6A

Solutions to exercises on page 47

CONVENTIONAL REPRESENTATION

1	7C
2	7A
3	3A
4	8C
5	5A
6	3C
7	5B
8	1A
9	3B
10	5C

Solutions to exercise on page 48

1. Section line	2. Centre line	3. Rib	4. Boss
5. Foot (Base)			
6. Cutting Plane	7. Taper	8. Undercut	9. M/c'ing symbol
10. EXT thread			
11. Diamond knurl	12. Spotface	13. Hidden detail	14. Diameter
15. Fillet radius			
16. CSK hole	17. Millimetre	18. Third	19. Centres
20. A Bush			

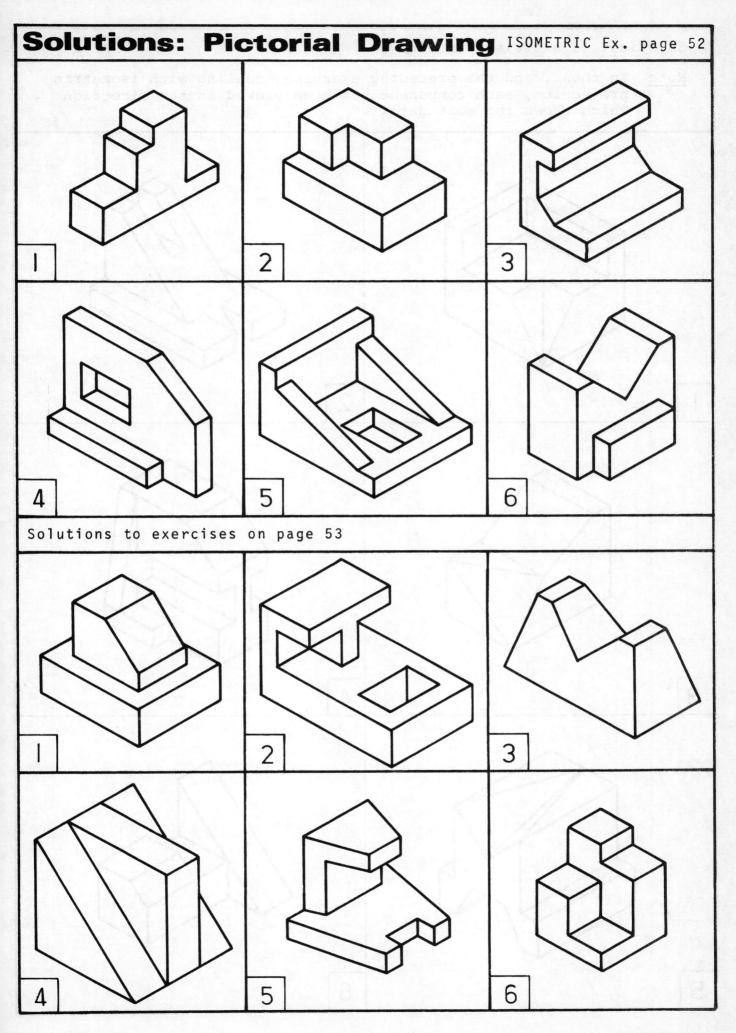

Solutions: Pictorial Drawing

ISOMETRIC Ex. page 52

Solutions to exercises on page 53

149

Note   In these, and the preceding exercises dealing with isometric
       projection, each component has been viewed in the direction
       which shows the most detail.

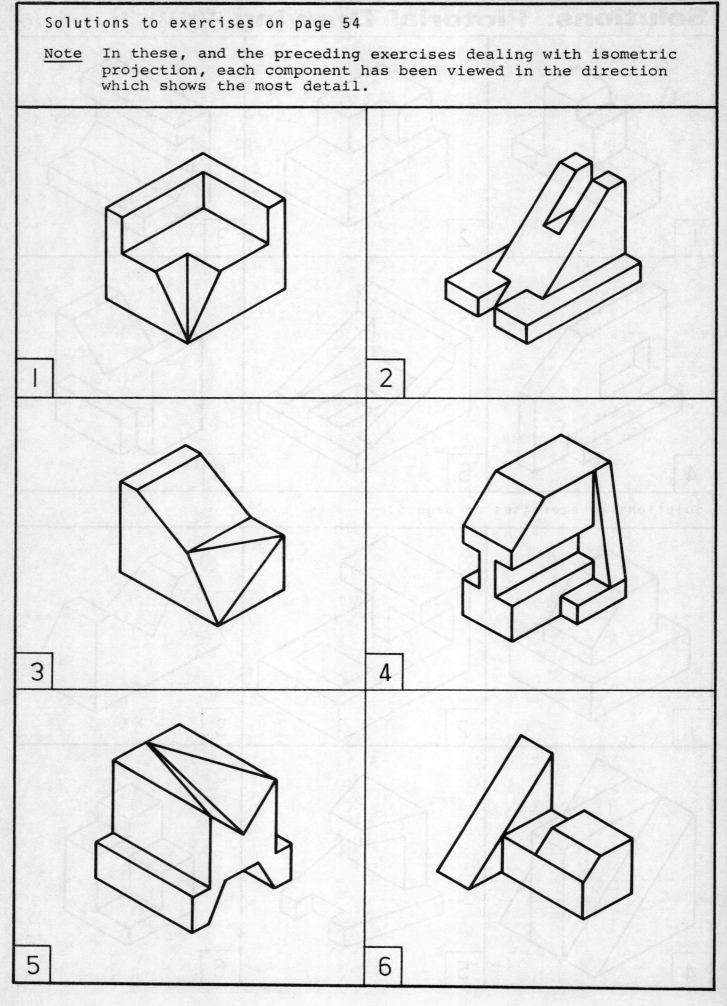

1

2

3

4

5

6

Solutions to exercises on page 58

<u>Note</u> Each view has been drawn looking the directions indicated in the exercise. These were chosen so as to show the most detail in the pictorial view. All solutions have been drawn using either an isometric grid or a framework of lines as in 2.

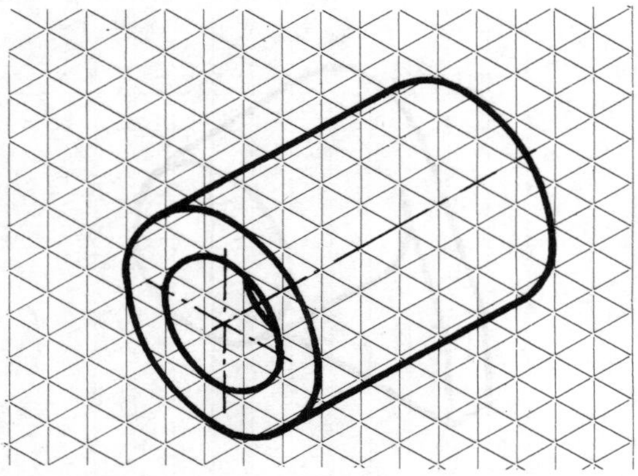

1

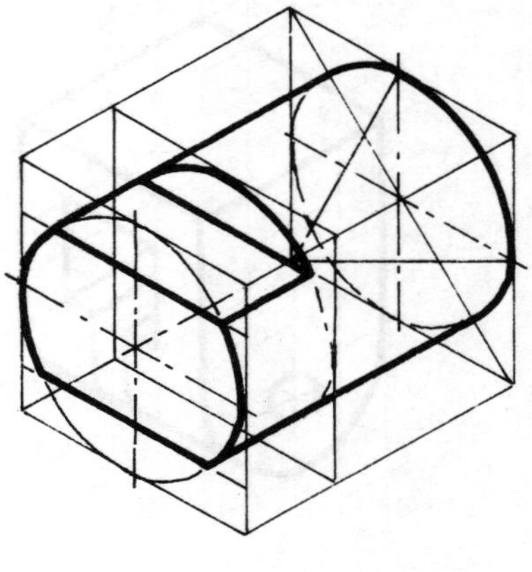

2

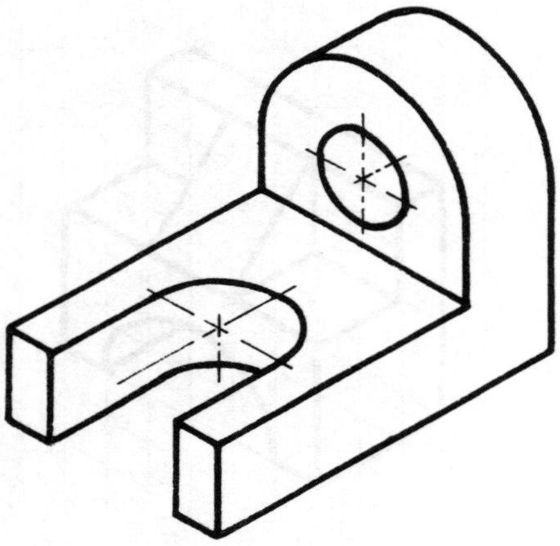

3

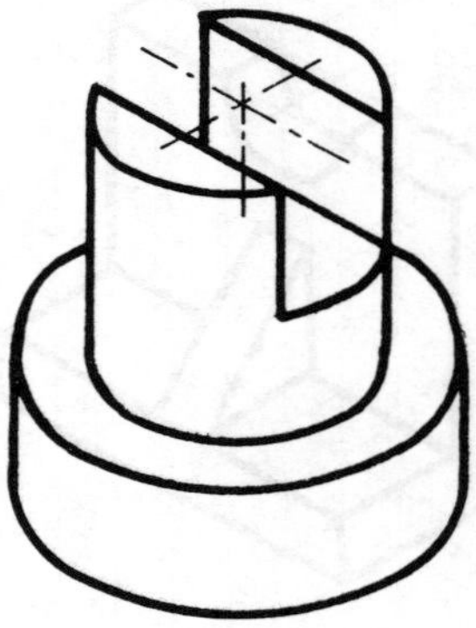

4

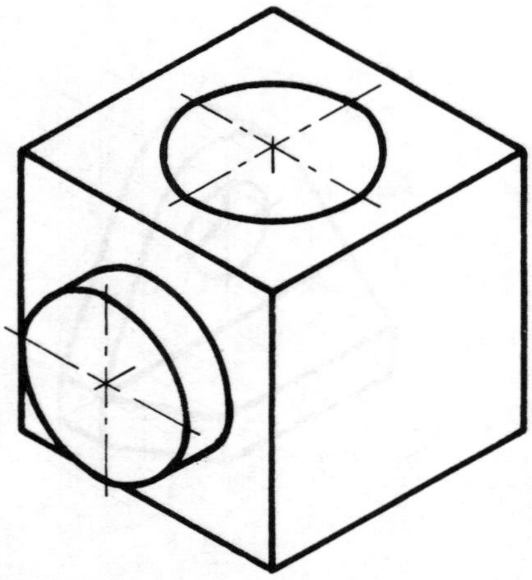

5

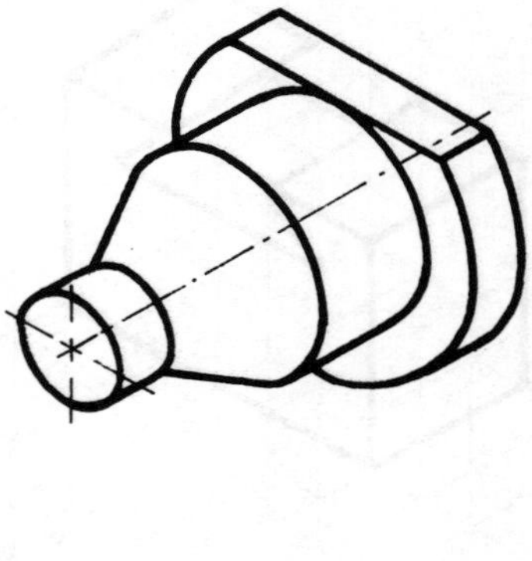

6

Solutions to exercises on page 59

Note   Each solution has been sketched freehand on an isometric grid.
       The resulting pictorial views are those which show the most
       detail of each component.

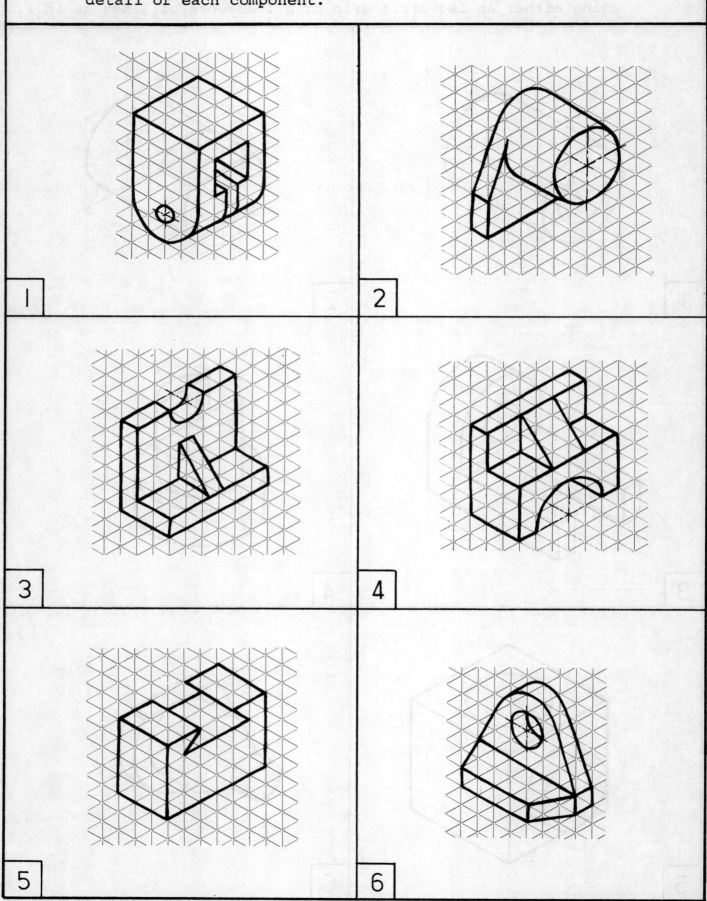

152

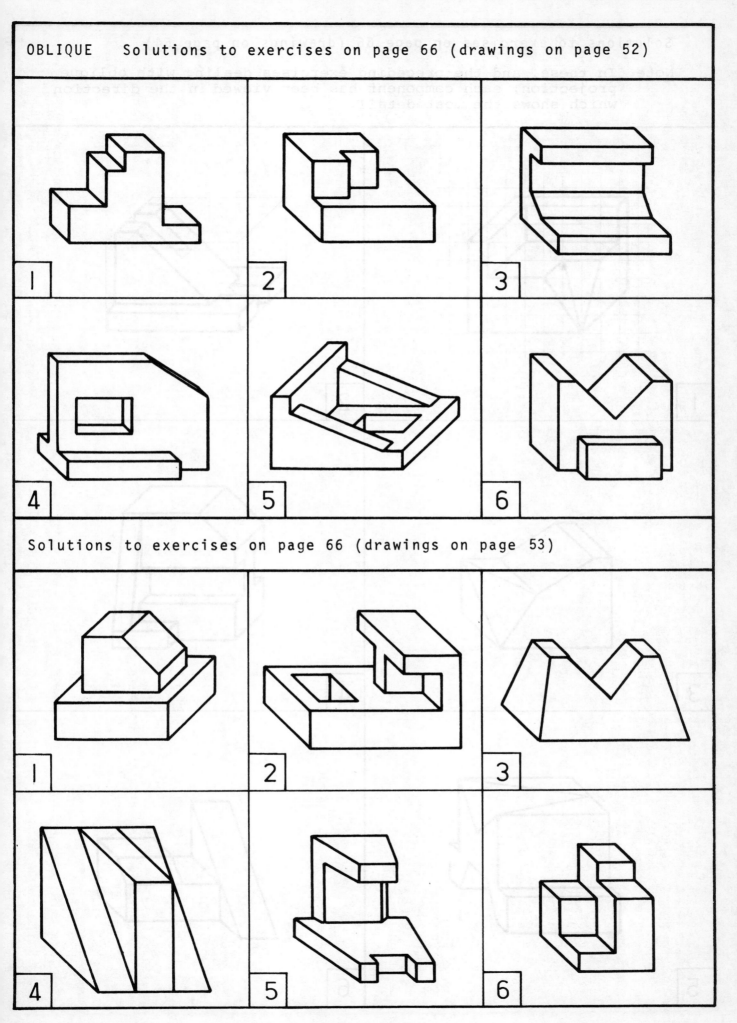

OBLIQUE   Solutions to exercises on page 66 (drawings on page 52)

1

2

3

4

5

6

Solutions to exercises on page 66 (drawings on page 53)

1

2

3

4

5

6

Note In these, and the preceding exercises dealing with oblique
projection, each component has been viewed in the direction
which shows the most detail.

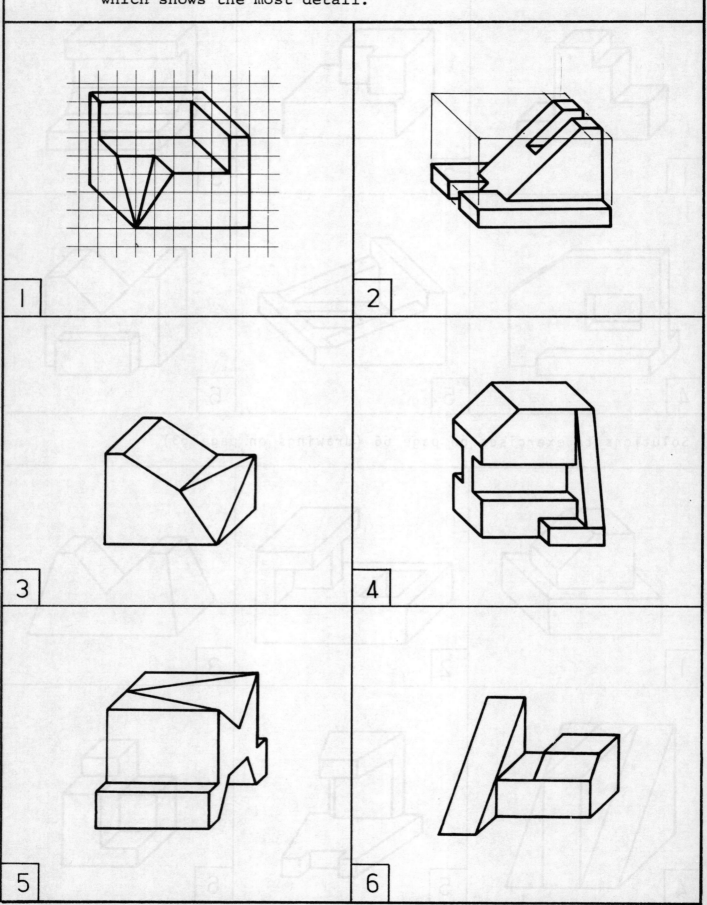

1

2

3

4

5

6

Solutions: Dimensioning

Solutions to
exercises on page 72

74

and 75

Note

The solutions given are
ideally spaced because
there is adequate space
for dimension lines
and numbers.
On an engineering drawing,
other lines may reduce
the amount of space
available for dimensions
and care must be taken to
ensure that lines and
numbers are clear and do
not clash with any other
detail.
This is particularly the
case with circles and
radii when the dimension
is often taken outside
the circle or the radius.

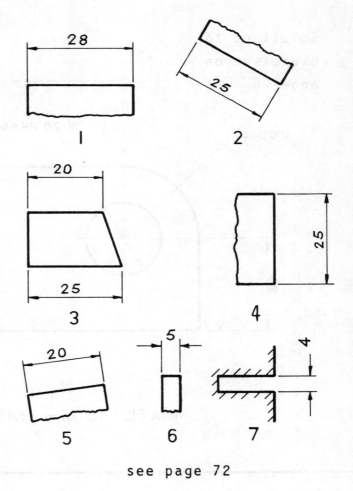

see page 72

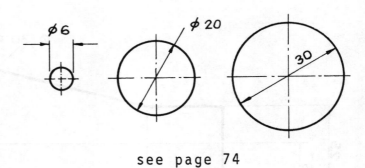

see page 74

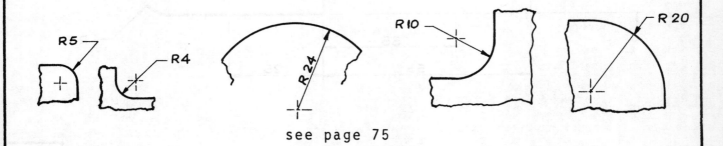

see page 75

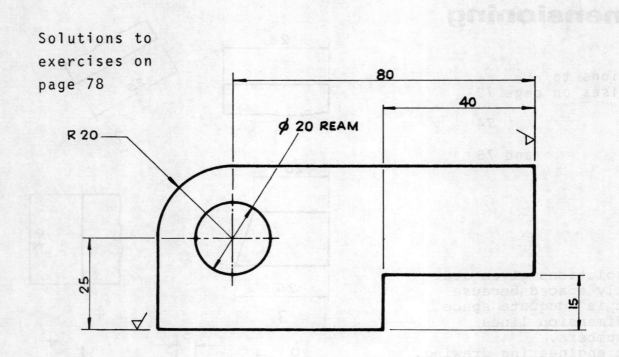

MATL - 15 mm MILD STEEL.

1

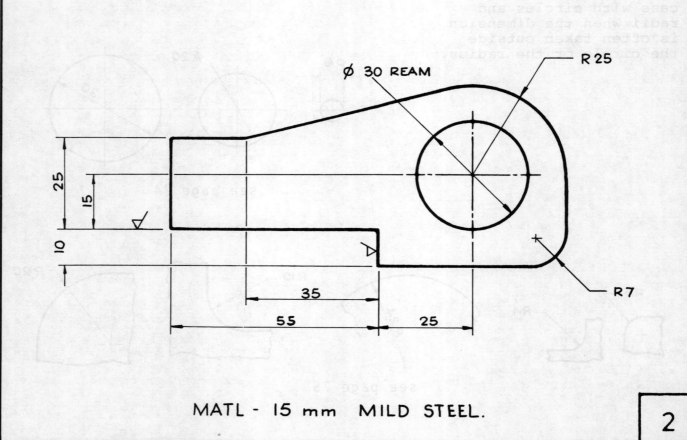

MATL - 15 mm MILD STEEL.

2

Solution to exercise
on page 79

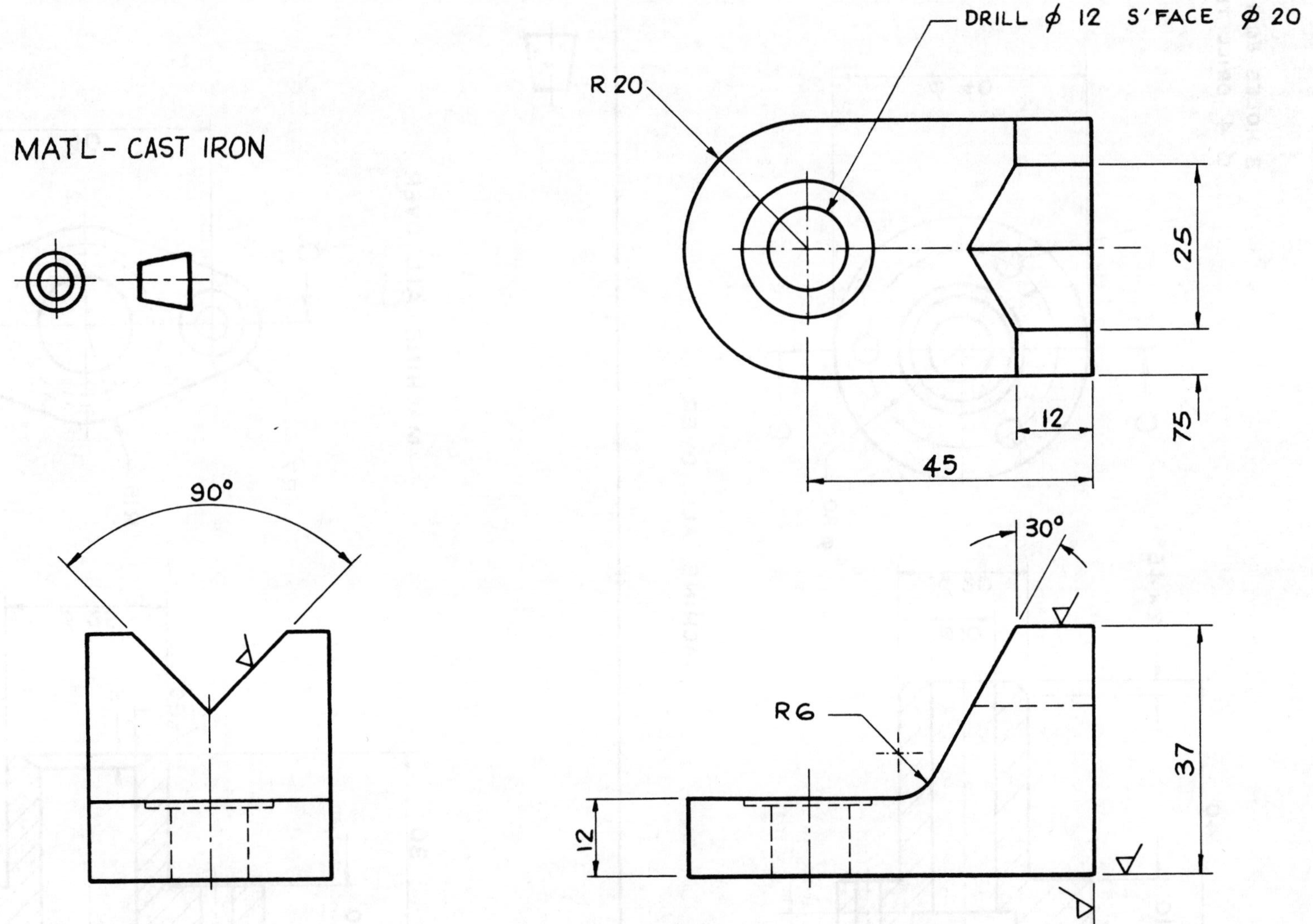

MATL – CAST IRON

DRILL ⌀ 12 S'FACE ⌀ 20

R 20

25

75

12

45

37

12

30°

R6

90°

3

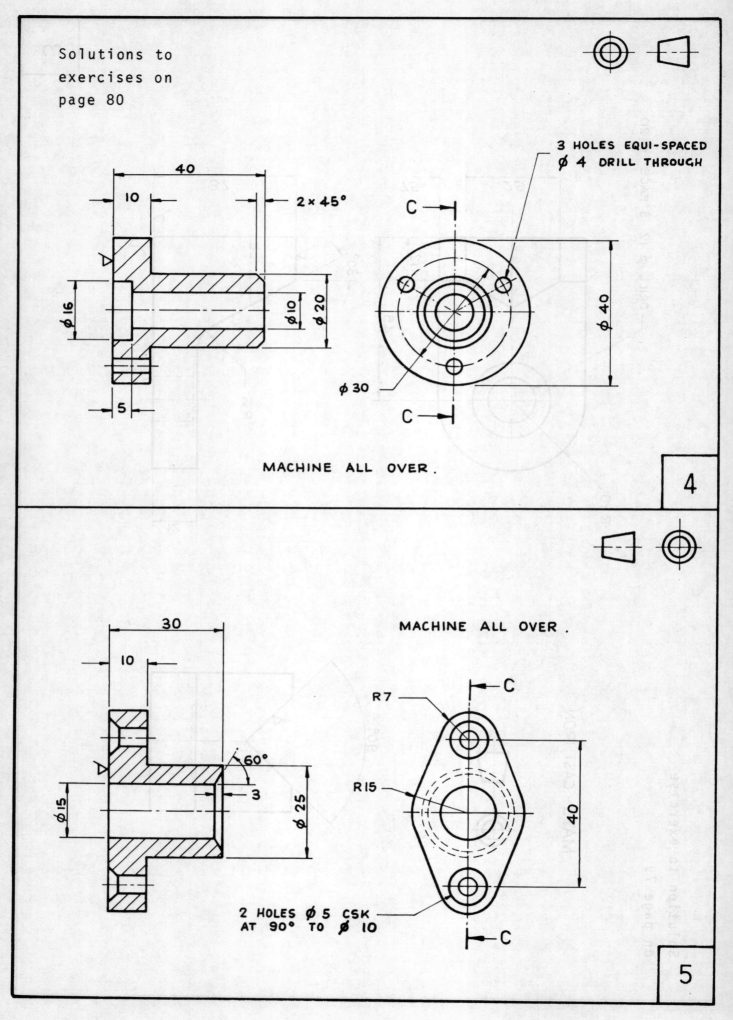

Solutions to
exercises on
page 80

40

10

2 × 45°

Ø16

Ø10

Ø20

3 HOLES EQUI-SPACED
Ø 4 DRILL THROUGH

C

Ø 40

Ø 30

C

5

MACHINE ALL OVER.

4

MACHINE ALL OVER.

30

10

60°

3

Ø15

Ø 25

R7

R15

C

40

2 HOLES Ø 5 CSK
AT 90° TO Ø 10

C

5

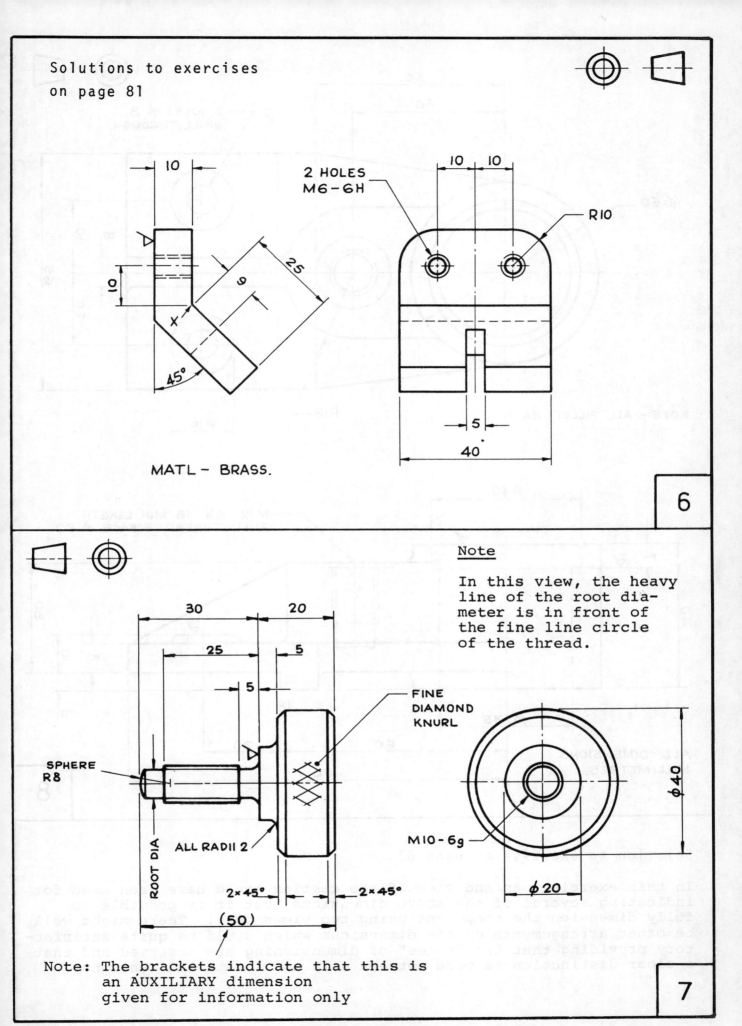

2 HOLES
M6-6H

R10

10

10 10

45°

25

9

X

MATL - BRASS.

5

40

6

Note

In this view, the heavy
line of the root dia-
meter is in front of
the fine line circle
of the thread.

30

20

25

5

5

FINE
DIAMOND
KNURL

SPHERE
R8

ROOT DIA

ALL RADII 2

2×45°

2×45°

(50)

M10-6g

φ20

φ40

Note: The brackets indicate that this is
an AUXILIARY dimension
given for information only

7

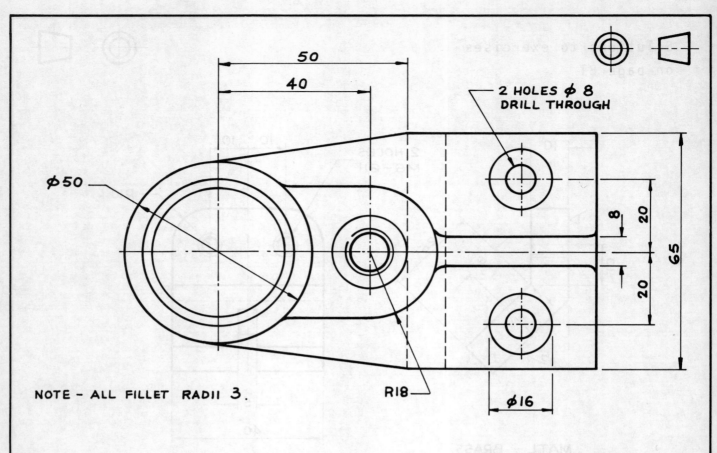

50

40

Ø50

2 HOLES Ø 8
DRILL THROUGH

8   20

65

20

Ø16

NOTE – ALL FILLET RADII 3.

R18

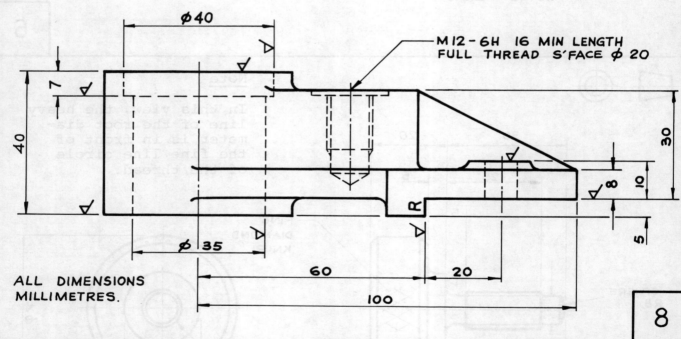

Ø40

M12 – 6H 16 MIN LENGTH
FULL THREAD S'FACE Ø 20

7

40

30

10

8

5

Ø 35

R

60

20

ALL DIMENSIONS
MILLIMETRES.

100

8

Solution to exercise on page 82

In this exercise, an end view of the casting could have been used for
indicating several of the above dimensions, but it is possible to
fully dimension the component using two views only. There might well
be other arrangements of the dimensions which would be quite satisfactory providing that the "rules" of dimensioning are observed and that
a clear distinction is made between size and location dimensions.

Solutions: Limits and Fits

Specimen solutions to exercises on page 90

Note: All sizes are in millimetres.
        The decimal point is on the base line.

H7-g6

	Hole				Shaft				
Basic Size	ES +	EI	Max. Size	Min. Size	Basic Size	es −	ei −	Max. Size	Min. Size
10	0.015	0	10.015	10.000	10	0.005	0.014	9.995	9.986
15	0.018	0	15.018	15.000	15	0.006	0.017	14.994	14.983
25	0.021	0	25.021	25.000	25	0.007	0.020	24.993	24.980
40	0.025	0	40.025	40.000	40	0.009	0.025	39.991	39.975
65	0.030	0	65.030	65.000	65	0.010	0.029	64.990	64.971

H7-s6

	Hole				Shaft				
Basic Size	ES +	EI	Max. Size	Min. Size	Basic Size	es +	ei +	Max. Size	Min. Size
8	0.015	0	8.015	8.000	8	0.032	0.023	8.032	8.023
18	0.018	0	18.018	18.000	18	0.039	0.028	18.039	18.028
28	0.021	0	28.021	28.000	28	0.048	0.035	28.048	28.035
38	0.025	0	38.025	38.000	38	0.059	0.043	38.059	38.043
78	0.030	0	78.030	78.000	78	0.078	0.050	78.078	78.050

H7-k6

	Hole				Shaft				
Basic Size	ES +	EI	Max. Size	Min. Size	Basic Size	es +	ei +	Max. Size	Min. Size
7	0.015	0	7.015	7.000	7	0.010	0.001	7.010	7.001
11	0.018	0	11.018	11.000	11	0.012	0.001	11.012	11.001
20	0.021	0	20.021	20.000	20	0.015	0.002	20.015	20.002
50	0.025	0	50.025	50.000	50	0.018	0.002	50.018	50.002
70	0.030	0	70.030	70.000	70	0.021	0.002	70.021	70.002

Solutions: Sectioned Assemblies

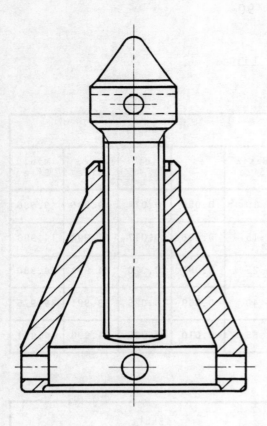

Exercises on pages 99 and 100

On page 99

(1) Sectioned assembly looking on
    cutting plane CC.

<u>Note</u>:

Although the screw lies in the
cutting plane it is <u>not</u> sectioned
when cut longitudinally
(i.e. along its length).

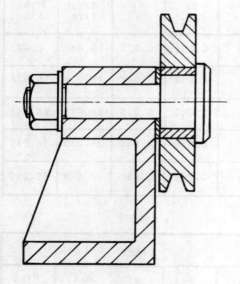

On page 99

(2) Sectioned assembly looking on
    cutting plane CC.

<u>Note</u>:

The pin, nut and washer are
<u>not</u> sectioned.
The spacer is not considered
to be the same as a washer
and is sectioned in this case.

On page 100

(3) Sectioned assembly looking on
    cutting plane CC.

<u>Note</u>:

The hinge pin lies across the
cutting plane and <u>is</u> sectioned
when cut transversely.

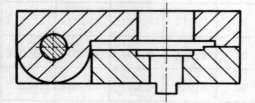

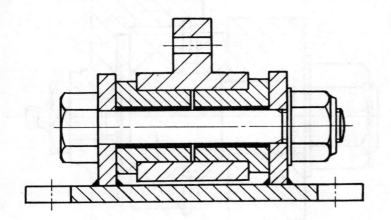

On page 100

(4) Sectioned assembly looking on cutting plane CC.

<u>Note:</u>

The thin steel liners are sectioned by single thick lines

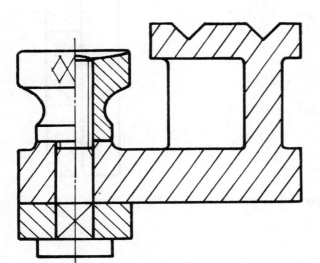

On page 101

(5) Sectioned assembly looking on cutting plane CC.

<u>Note:</u>

As the Vee grooves are at 45°, the section lines may be drawn at an angle other than the normal 45°.

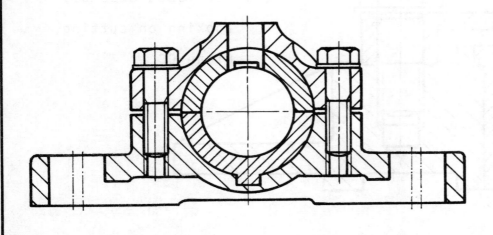

On page 102

(6) Sectioned assembly looking on cutting plane CC.

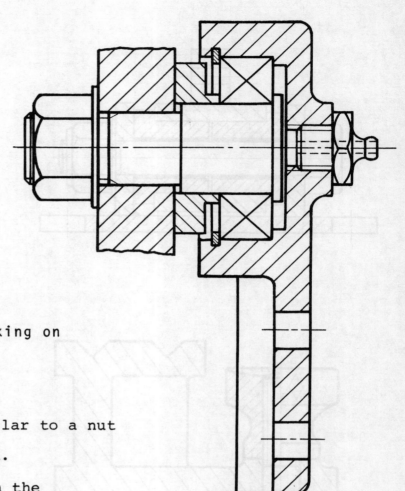

On page 103

(7) Sectioned Assembly looking on

cutting plane CC.

<u>Note</u>:

The grease nipple is similar to a nut

and need not be sectioned.

The ball race is shown in the

conventional manner.   See page 41.

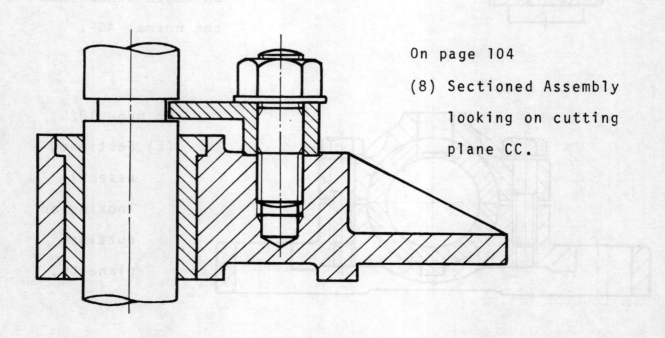

On page 104

(8) Sectioned Assembly

looking on cutting

plane CC.

Solutions: Developments: Parallel Line

In the solutions of development exercises, the following notes will be useful:

(1) Types of line.   Outlines   ——————————

                      Bend lines   — — — — — —

                      Construction lines   ————————

(2) All the prisms, cylinders, pyramids, cones and transition pieces are hollow, thus no section lines are required on cut surfaces.

(3) The development of a sheet metal component can be shown to be correct by forming the required shape from the pattern. A tracing can be made of each solution and transferred to thin card from which the pattern can be cut. An allowance must be made for extra material along fixing edges. See the example on page 105. The card folds more easily if the bend lines are scored on the inside surface of the pattern with a fine ball point pen. A quick drying latex type adhesive is ideal for joints.

PARALLEL LINE DEVELOPMENT

Solutions to exercises on page 107

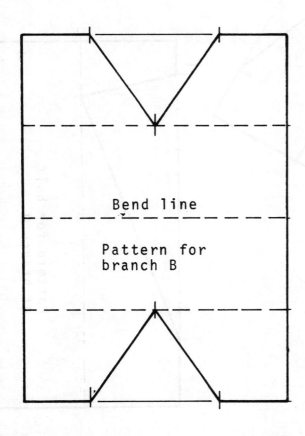

Bend line

Pattern for branch B

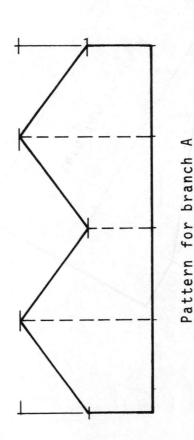

Pattern for branch A

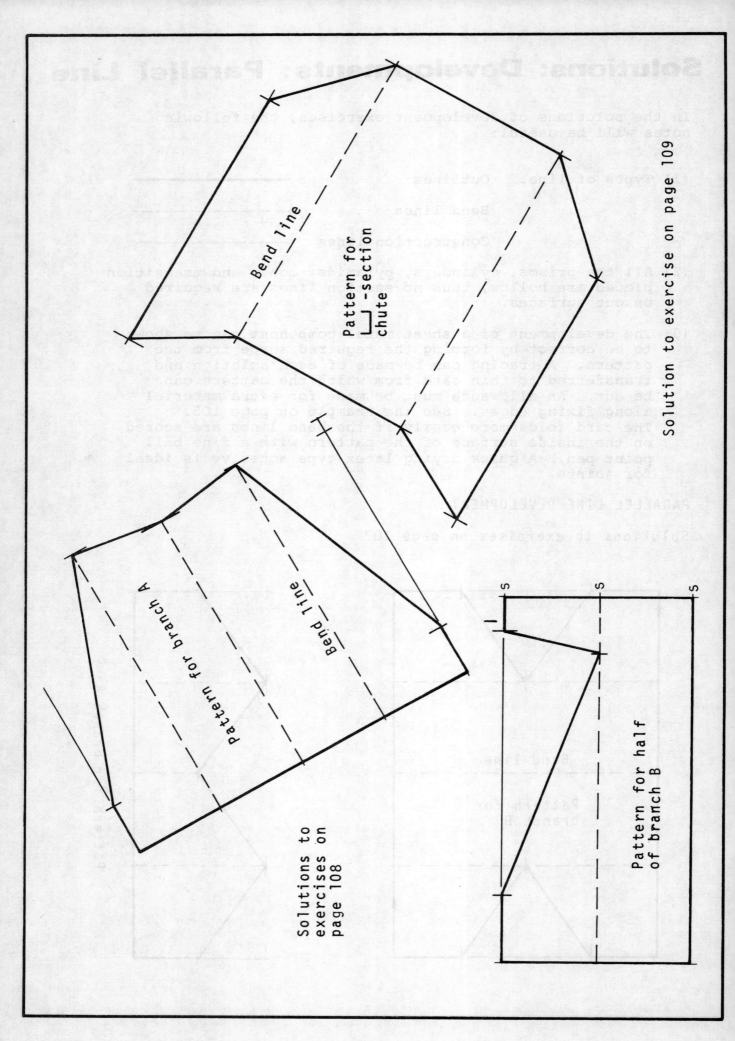

Bend line

Pattern for ⌐section chute

Solution to exercise on page 109

Pattern for branch A

Bend line

Pattern for half of branch B

Solutions to exercises on page 108

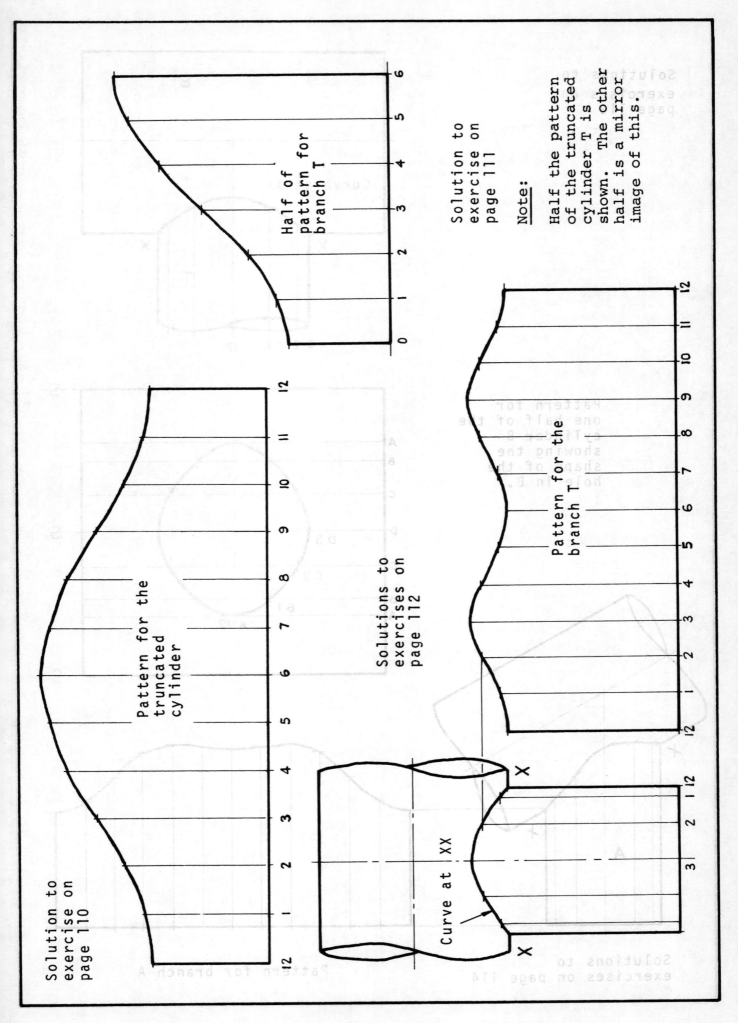

Solution to exercise on page 110

Pattern for the truncated cylinder

Half of pattern for branch T

Solution to exercise on page 111

Note:

Half the pattern of the truncated cylinder T is shown. The other half is a mirror image of this.

Solutions to exercises on page 112

Pattern for the branch T

Curve at XX

Solutions to exercises on page 114

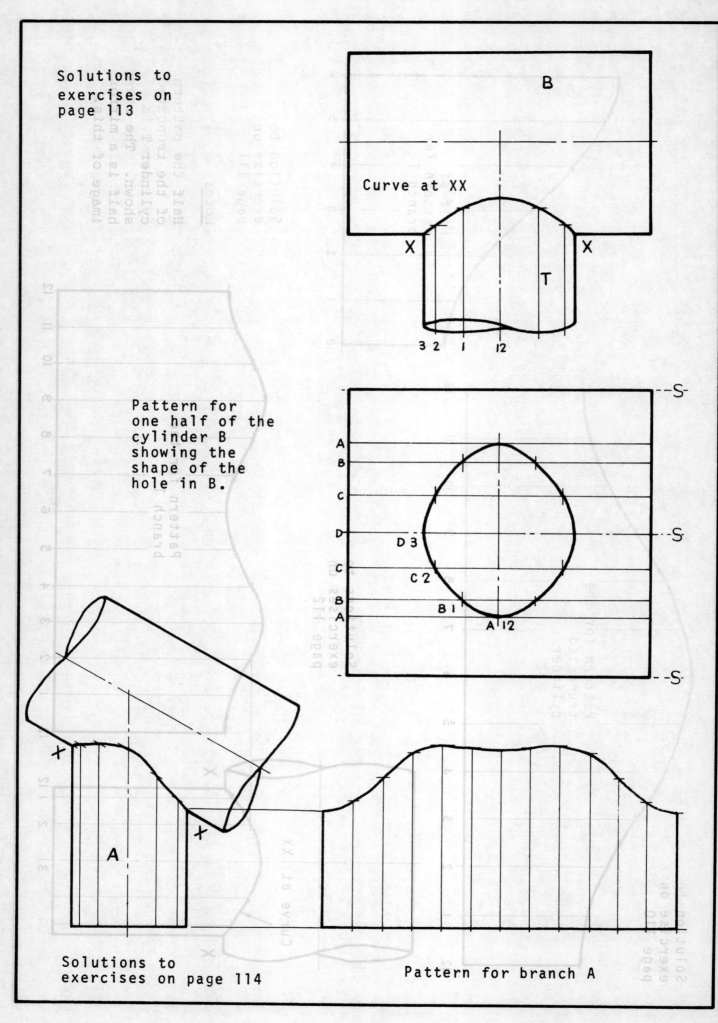

B

Curve at XX

X   X

T

3 2 1   12

Pattern for
one half of the
cylinder B
showing the
shape of the
hole in B.

-S-

A
B
C

D   D 3   S

C   C 2

B   B 1
A   A 12

-S-

A

X   X

Pattern for branch A

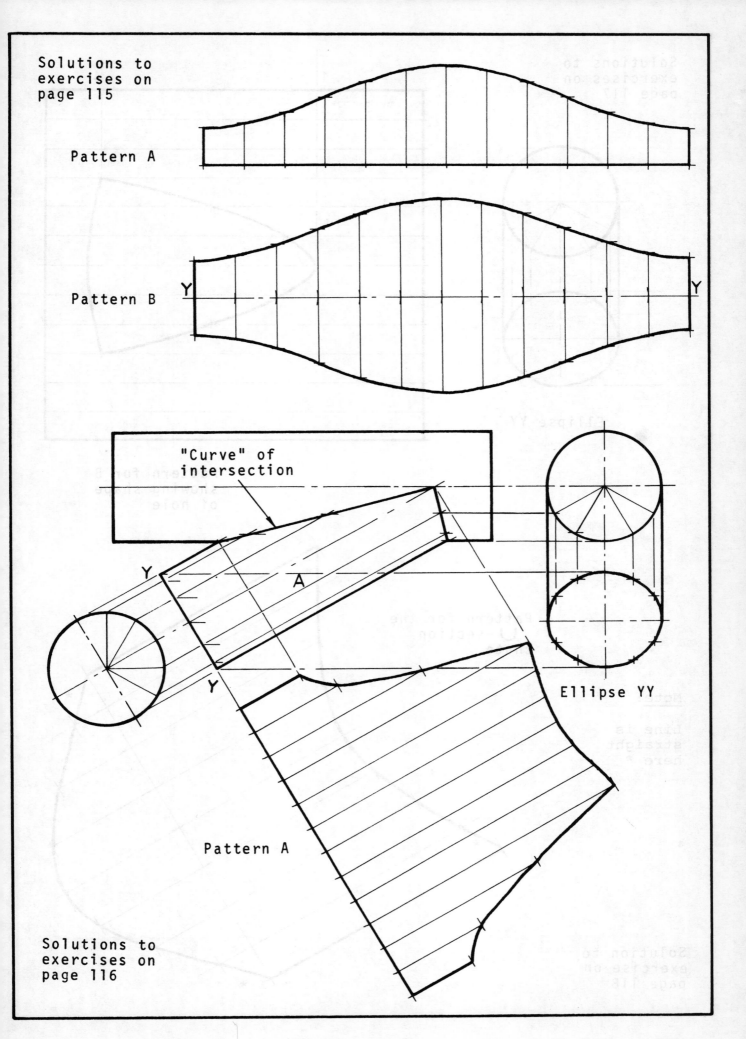

Solutions to
exercises on
page 115

Pattern A

Pattern B

Y   Y

"Curve" of
intersection

Y

A

Y

Ellipse YY

Pattern A

Solutions to
exercises on
page 116

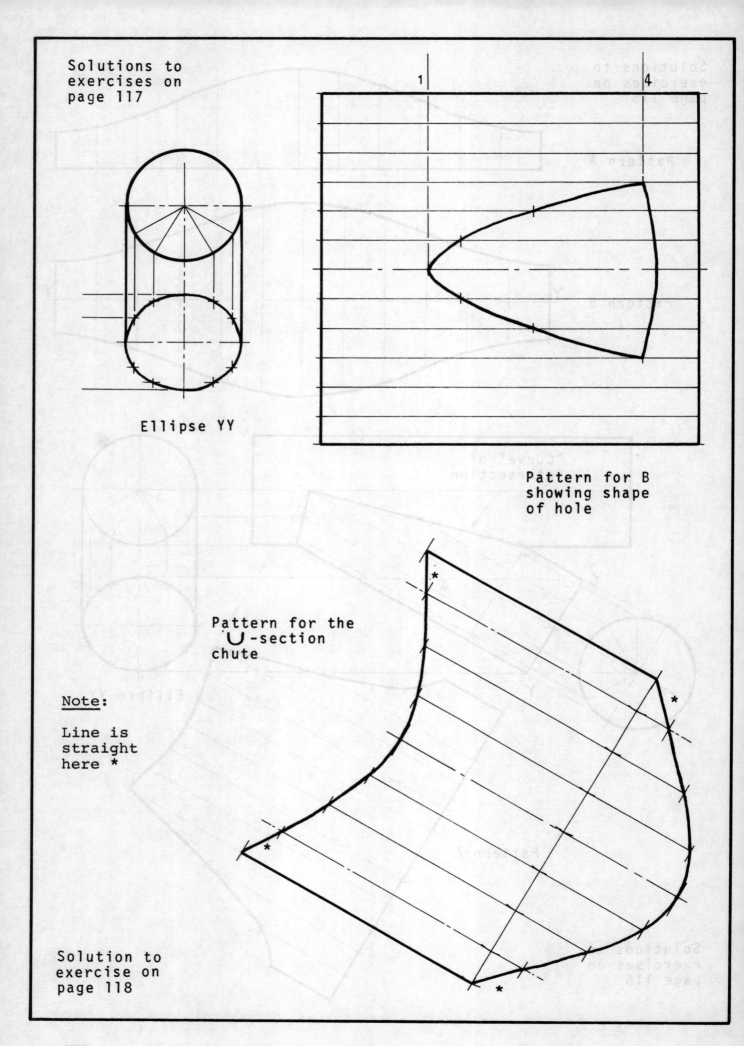

Solutions to
exercises on
page 117

1　　　　　　　　4

Ellipse YY

Pattern for B
showing shape
of hole

Pattern for the
∪-section
chute

Note:

Line is
straight
here *

Solution to
exercise on
page 118

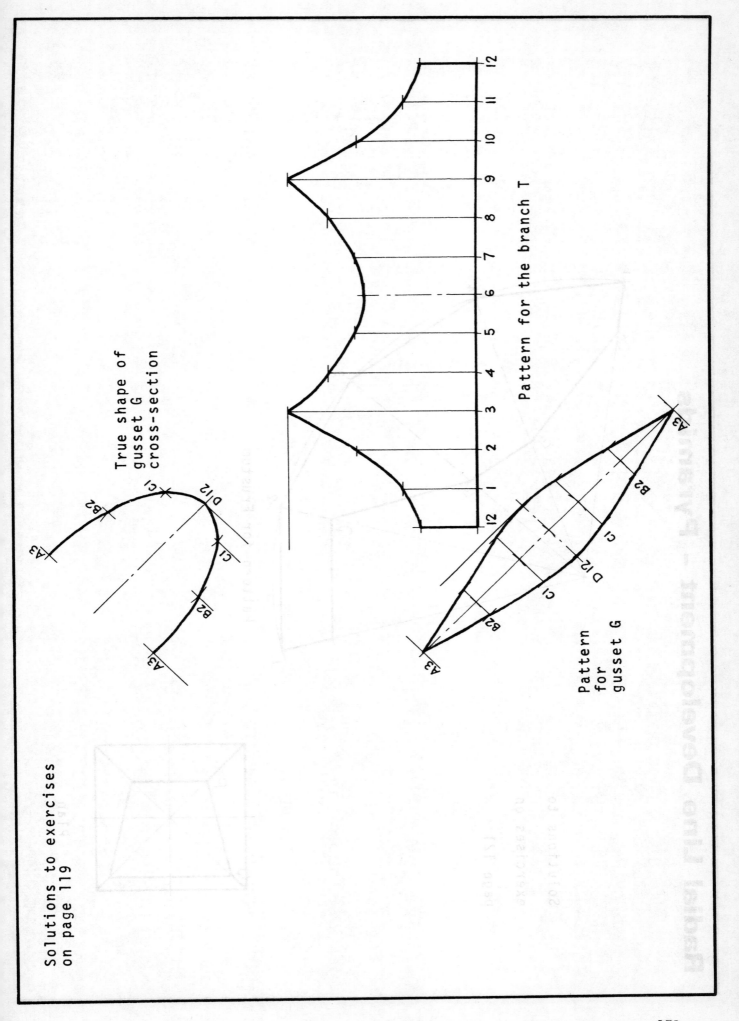

Solutions to exercises
on page 119

True shape of
gusset G
cross-section

Pattern for the branch T

Pattern
for
gusset G

Radial Line Development – Pyramids

Solutions to
exercises on
page 121

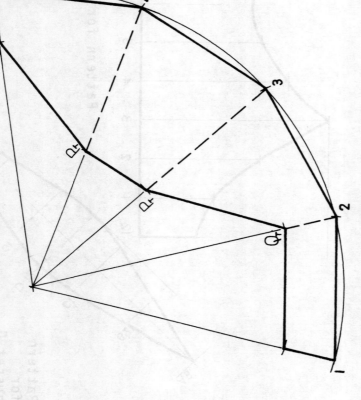

Pattern for Frustum

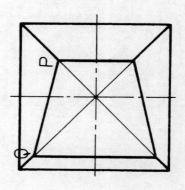

Plan

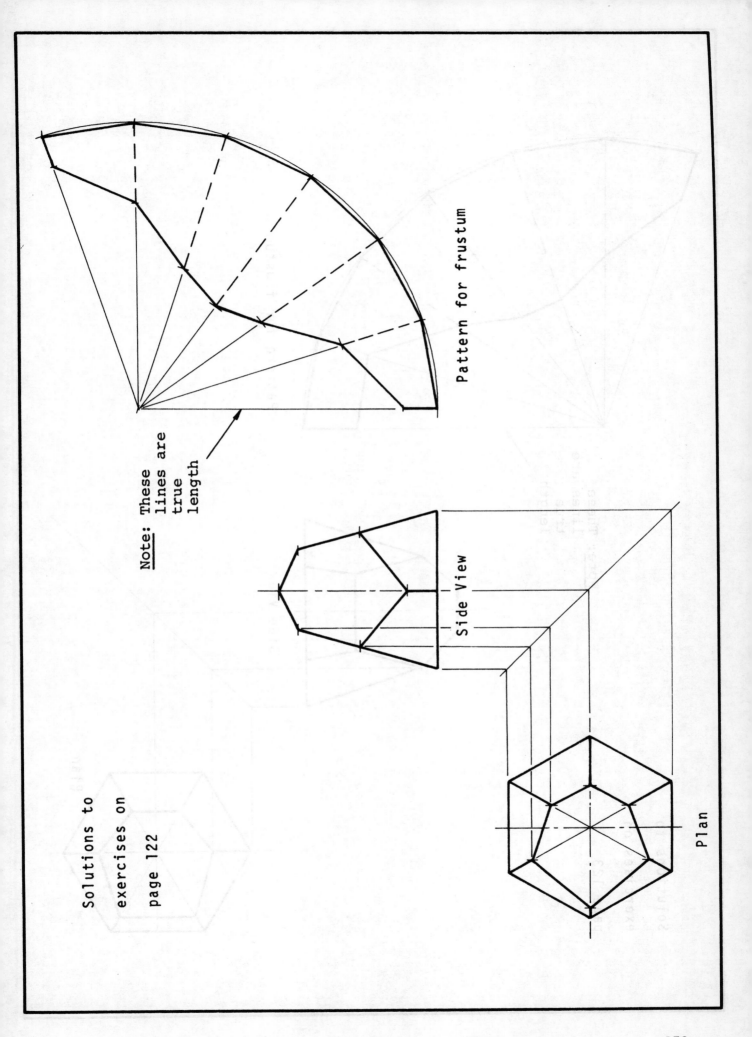

Pattern for frustum

Note: These lines are true length

Side View

Solutions to exercises on page 122

Plan

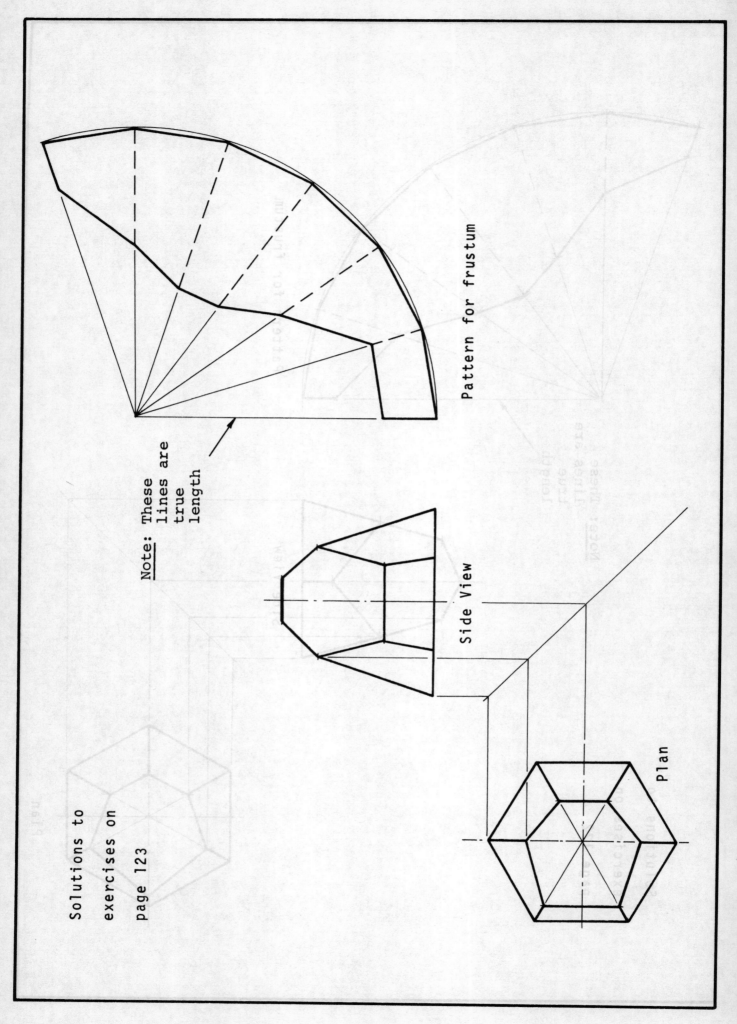

Pattern for frustum

Note: These lines are true length

Side View

Plan

Solutions to exercises on page 123

174

Radial Line Development – Cones

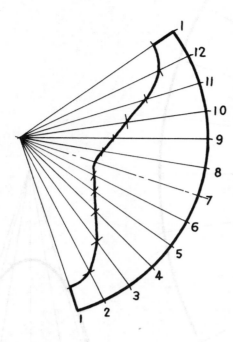

Complete pattern for
surface area of frustum
of a right cone

Completion of exercises
started on page 125

Note:

A RIGHT CONE is one in which
a line drawn from the apex of
the cone to the centre of the
circular base is always at
right angles to the base.

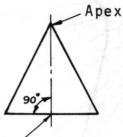

An OBLIQUE CONE is one in which
the line from apex to the centre
of the circular base is inclined
at any angle other than 90°
to the base.

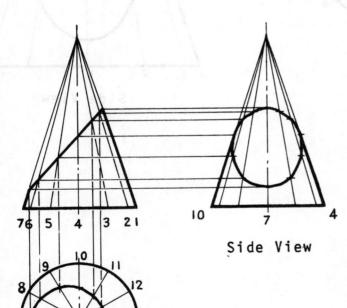

Side View

Plan

Note:

Section lines are not drawn

on cut surfaces because

the cone is hollow

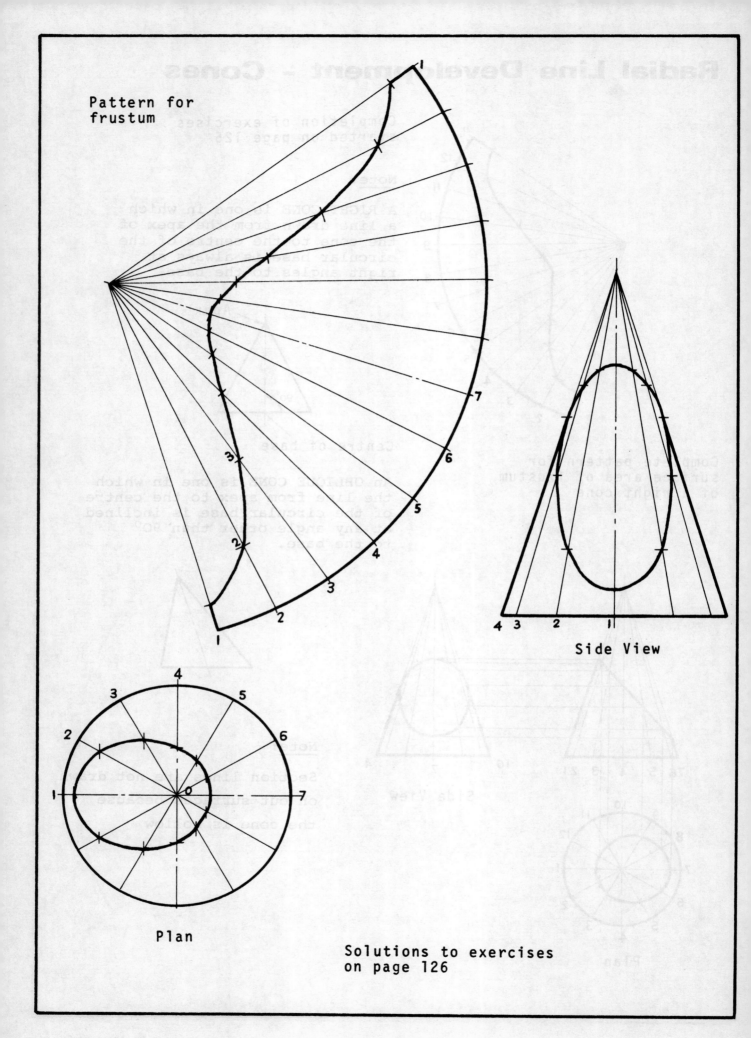

Pattern for
frustum

1

7

3

6

2

5

3

4

2

3

1

Side View

4 3 2 1

Plan

4

3

5

2

6

1

O

7

Solutions to exercises
on page 126

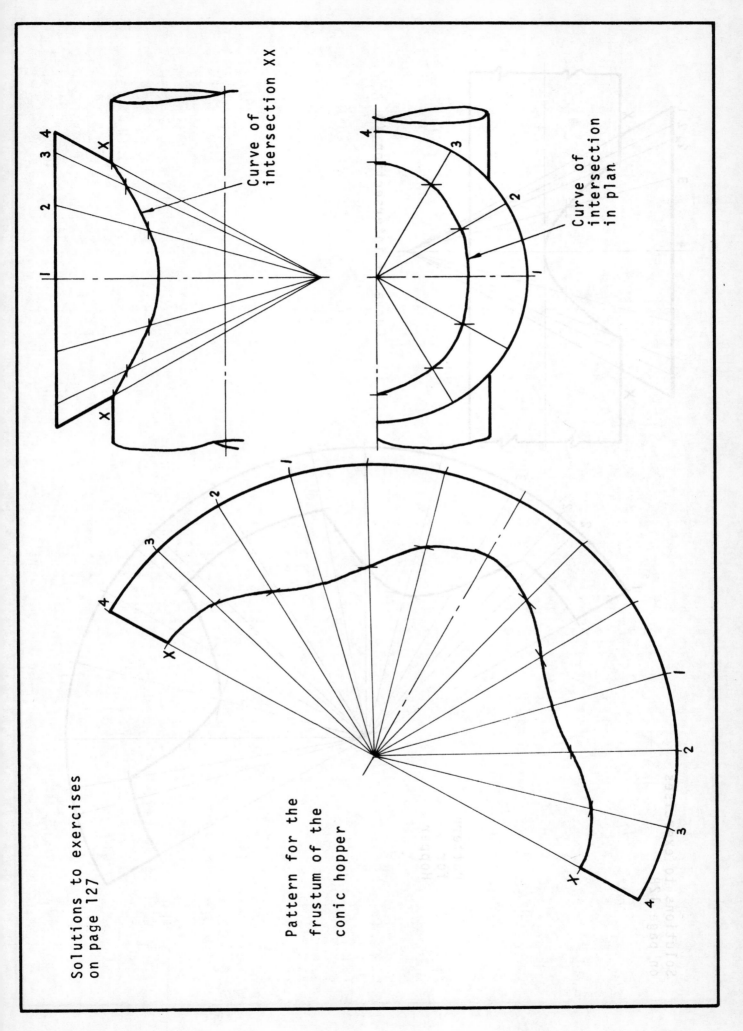

Solutions to exercises
on page 127

Pattern for the
frustum of the
conic hopper

Curve of intersection XX

Curve of
intersection
in plan

177

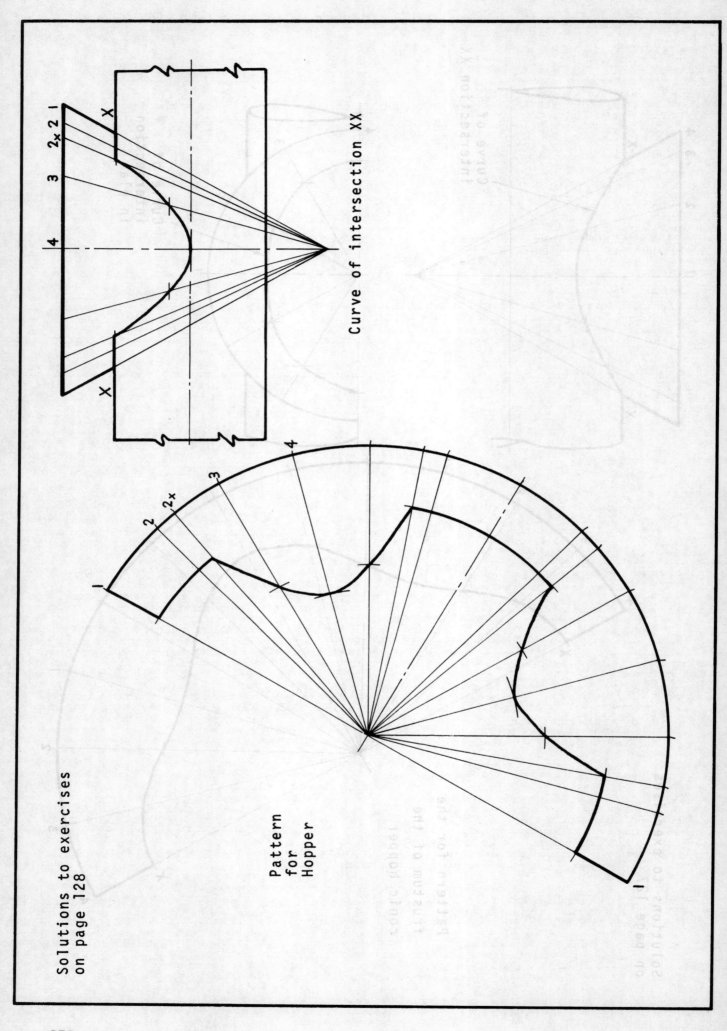

Solutions to exercises
on page 128

Pattern
for
Hopper

Curve of intersection XX

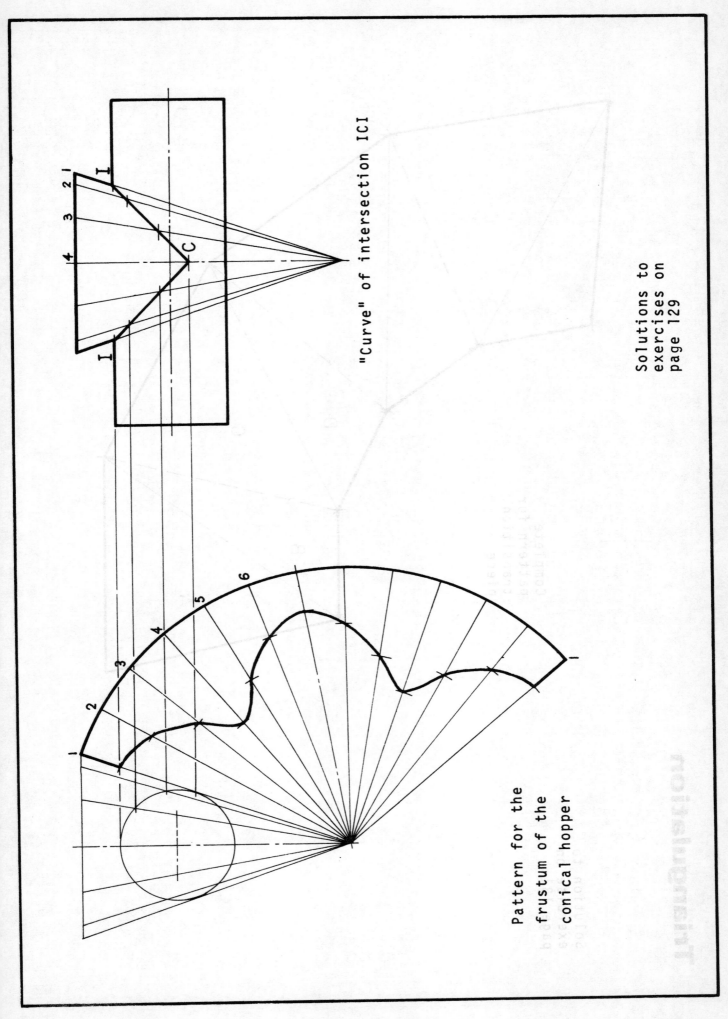

"Curve" of intersection ICI

Solutions to
exercises on
page 129

Pattern for the
frustum of the
conical hopper

Triangulation

Solution to
exercise on
page 131

Complete
pattern for
transition
piece

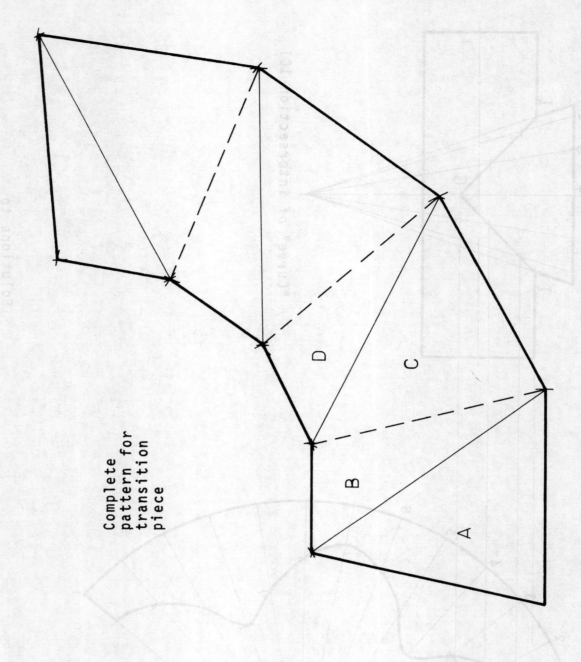

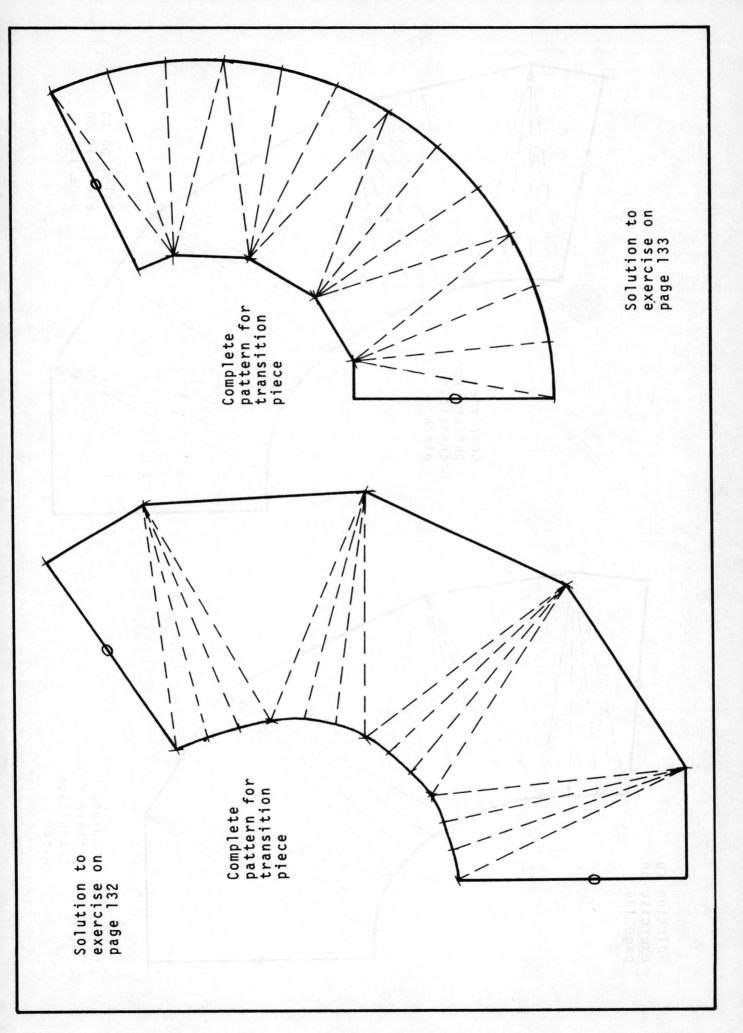

Solution to
exercise on
page 132

Complete
pattern for
transition
piece

Complete
pattern for
transition
piece

Solution to
exercise on
page 133

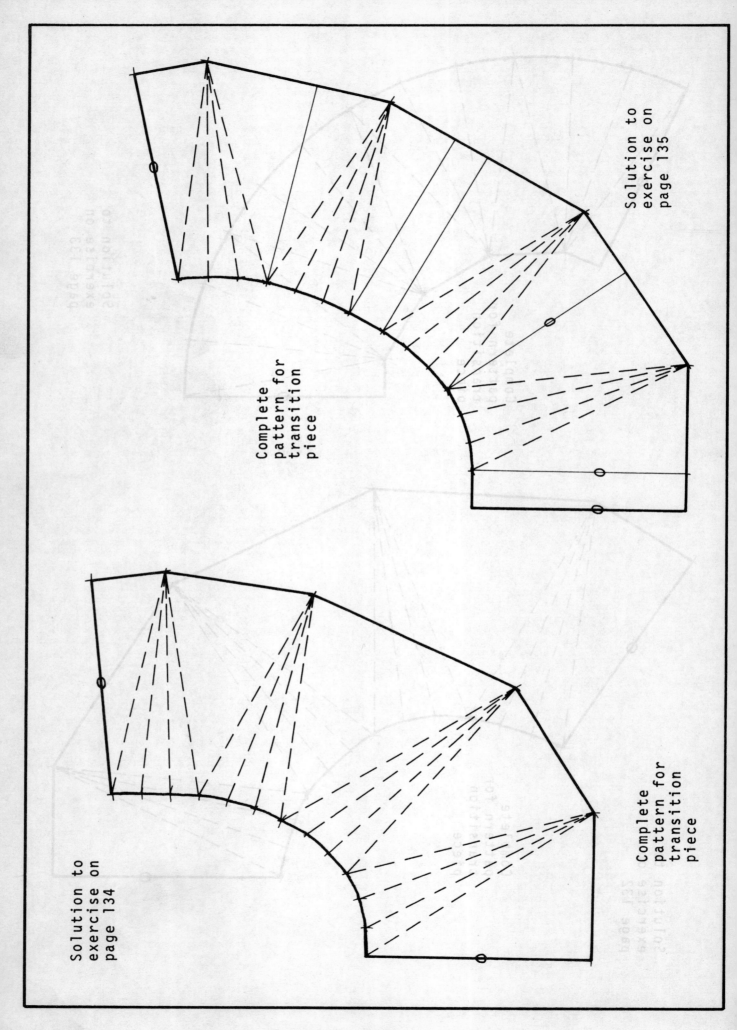

Solution to
exercise on
page 135

Complete
pattern for
transition
piece

Solution to
exercise on
page 134

Complete
pattern for
transition
piece